Who Do We Think We Are?

Lord's blessings

Herb
Dec. 13, 2014

"I love Gruning's humorous, engaging style and his ability to take seriously and integrate religious traditions, philosophy, and a wide variety of scientific research in his quest to understand reality, without letting any ideas go unchallenged."
—Paul van Arragon, computer scientist

"This book puts into question one's own habits of thinking regarding our 'human' nature. 'We never grow too old to construct a different us.'"
—Maurice Boutin, John W. McConnell Emeritus Professor of Philosophy of Religion, McGill University, Montreal, Canada

"The question of the ages: Who Do We Think We Are? Before the Enlightenment this question, at least for European civilization, was literally immaterial. Since the 18th century, however, the approach has been one of engineering the perfect human being, and thereby the perfect human society. In his fourth book, Dr Gruning examines this vexing problem from historical, biblical and personal perspectives for an intriguing and unique assessment that gauges the heights of the human spirit, while casting a pessimistic note on our prospects."
—Keith Sudds, retired industrial worker and student of the human condition.

"Like his earlier works, Dr Herb Gruning's latest book examining what it means to be human covers subject matter of keen personal interest . The subject is far from my areas of technical expertise, and I greatly appreciate his clear and succinct identification of different theorists and schools of thought, without the excess detail that would make this highly readable book into an exhausting tome. Written in relaxed style, this conversation with a friend has challenged and informed, prompting arresting inner dialogue between my mind and soul or mind and spirit, as well as external conversation with the divine."
—Ian Moore, PhD, PEng, Professor of Civil Engineering, Queen's University at Kingston

"Can we know who we are? Herb Gruning attacks this ageless mystery playfully but extensively, through many cultural lenses. He enjoys that God is a step ahead of our most scholarly wonderings."
—Carolyn Smith, Minister of Christian Development and Pastoral Care, Applewood United Church

"In *Who Do We Think We Are?*, Herb Gruning writes about big ideas in an accessible, reader-friendly way. He does not propose a definitive answer to questions but solicits the reader's participation in a reflective journey through theological and philosophical issues vital to human self-understanding. With refreshing honesty and an inimitable personal touch, he addresses perennial issues that deserve a hearing in any reflective, intelligent life.

"Trying to make sense of human nature is a formidable task; the problems the subject poses are multilayered and formidable. Gruning responds with a wide-ranging and meticulous treatment of seminal issues. He adopts a moderate, probing voice throughout. The book displays impressive historical scope, moving from the ancient to modern ideas to recent evolutionary and anthropological preoccupations. The author carefully investigates a multitude of religious and secular thinkers and panoply of relevant issues from biblical exegesis, to materialism, to idealism, to the mind-body problem, to reincarnation to pop-culture, to dreams.

"This book provides a stimulating if sometimes unsettling perspective on major themes. Religiously and philosophically inclined readers, those who like to go beyond clichés and canned stereotypes, will find the treatment absorbing. This is not philosophy for professors but philosophy for independent-minded readers who like to make up their minds on their own."
—Louis Groarke, (Full) Professor, Philosophy Department, St. Francis Xavier University, Antigonish, Nova Scotia

Copyright © 2015 Herb Gruning
All rights reserved.

Published by Blue Dolphin Publishing, Inc.
P.O. Box 8, Nevada City, CA 95959
Orders: 1-800-643-0765
Web: www.bluedolphinpublishing.com

ISBN: 978-1-57733-279-4 paperback
ISBN: 978-1-57733-443-9 e-book

Library of Congress Control Number: 2014951310

Scripture quotations are from the New Revised Standard Version Bible, copyright 1989, Division of Christian Education of the National Council of the Churches of Christ in the United States of America. Used by permission. All rights reserved.

Alternatively, scripture quotations marked (NIV) are taken from the Holy Bible, New International Version®, NIV®. Copyright © 1973, 1978, 1984, 2011 by Biblica, Inc.™ Used by permission of Zondervan. All rights reserved worldwide. www.zondervan.com The "NIV" and "New International Version" are trademarks registered in the United States Patent and Trademark Office by Biblica, Inc.™

Printed in the United States of America

10   9   8   7   6   5   4   3   2   1

*For Alice,
my favorite human*

Contents

Foreword by Kevin Krumrei — xi

Preface — xv

Part One: Biblical Material — 1

Part Two: Historical Material — 29

Part Three: Contemporary Material — 61

Part Four: Recent Developments — 93

Part Five: Case Studies — 145

Conclusion and Personal Reflections
(or Reflections on the Personal) — 169

Appendix: Dreamworld — 194

Bibliography — 199

Index — 205

About the Author — 214

Foreword

"What I am about to say is likely the most controversial thing I have ever said in print. The following subject matter contains violence. Reader discretion is advised," warns Dr. Herb Gruning. Those are fighting words, for sure. But those who are familiar with Gruning's work know that his project is one of reconciliation, not violence – bringing together what are often taken as contradictory points of view and finding a way to move forward. Gruning's process theology, his understanding of evolutionary theory, his approach to the Christian scriptures, and his metaphysics of science have come together to form a persuasively coherent view. This book is a solid addition to that legacy.

Let me rewind the clock a little. I met Herb Gruning when I was a fresh-faced graduate student in the Philosophy Department at Brock University. Gruning was, at the time, teaching a course there (World Religions, if I remember correctly). We hit it off immediately, having similar philosophical and theological interests, and he introduced me to a number of new ideas in the philosophy of religion and science. I remember pointing out, during one conversation, that some of his views (from process theology to his take on scripture) might not fit with classical Christian positions. "Well," he sighed, "I guess I'm just a liberal theologian."

Labels can be destructive. They are divisive, limiting generalizations that often do not help to move matters forward. I would not suggest that Gruning is either a "liberal" or a "conservative" theologian. What I would say is that those on the conservative end of the spectrum will find some of his positions unsettling, and so they should. His questions concerning the character and actions of God in the Old Testament are serious, sincere, and precisely the kind of questions that everyone interested in the Judeo-Christian tradition should wrestle with. Those on the liberal end might actually find his continued use of Christian theological concepts unsettling – Jesus as Messiah, the human predicament understood as "God-substitutes" (read: idolatry), the connection he draws between neuroplasticity and "the principalities and powers" (read: repentance), his use of the word "spirit", to name just a few. This is not the work of someone who is dismissive of religious traditions in the least. Gruning takes seriously biblical categories and shows how they can flourish within the most rigorous of scientific and philosophical frameworks.

Having said that, I would not call this a book of answers. Even the title is a question! But having completed three graduate degrees myself, I can appreciate the astonishing amount of knowledge required to even ask good and fruitful questions. Karl Popper (perhaps not Gruning's philosopher of choice) insisted that the way science moves forward is through falsification instead of verification, but this leads to provisional positions and queries, not foundational truths. Take Gruning's position on the mind/body problem (spoiler alert: he's not a materialist!). Here he stands firmly in the tradition of Sherlock Holmes, where whatever remains after all other possibilities have been eliminated must be the case, however unlikely. For him, dualism seems to be the last option still standing, which Gruning admits, leads only to mystifying questions. My point is this: I know of very few individuals who have the requisite expertise in evolutionary theory, metaphysics and the philosophy of science, and the history of religious thought and particularly biblical scholarship, to bring all of these viewpoints together and even begin considering the implications for theories of human nature in that interdisciplinary context. To accomplish this in a way that makes the reader feel educated rather than intimidated is quite simply a most impressive achievement.

Of particular interest, I think, are the following (the list is not exhaustive): Gruning's discussion of differences in degree and differences in kind with regard to human uniqueness strikes me as crucial to any theology of human nature that wishes to remain relevant within the broader scientific and cultural dialogue. His critical reflection on theology (and theologians), and also on the limits of scientific inquiry, are integral to his criticism of fundamentalism and dogma, both religious and scientific, and entirely apropos in the current academic climate. I have, for instance, not often heard someone first reject Intelligent Design theory, and then defend it from the "scientific inquisition" (usually an author does one or the other). His extension of Barbour's fourfold typology of the relation of science and religion to include a fifth, namely, mysticism, is a most intriguing notion. Also, his considerations of the mind/body problem, the possibility of an afterlife, and the implications therein for human nature in light of both scientific and religious considerations, are most enlightening.

I should also like to note that anyone who has spent an evening with Herb and his wife Alice will recognize quite clearly, chapter by chapter (if not page by page), the influence of his sense of humour which, on the dryness scale, falls somewhere between the arctic tundra and a Saharan sand dune. It is not often that I chuckle out loud while reading a book of this nature, and it did result in a number of quizzical and accusatory glances in the "quiet zone" on the Go-Train (a commuter train in the Greater Toronto Area), where it is undoubtedly most uncouth to laugh.

This book is a pleasure to read, and will challenge its readers, no matter what their philosophical or religious bent. There is something in this book to delight and trouble everyone.

Kevin Krumrei
Records and Information Technician, York Region

Preface

THIS BOOK, now my third with Blue Dolphin, is about human nature. The personal reflections contained herein take on a similar format as my previous volume, *God Only Knows*, also published by Blue Dolphin. Selected passages from the Judeo-Christian scriptures dealing with the topic are addressed and receive commentary, and this comprises Part One of our present study. Part Two examines several religious and philosophical systems to determine what "the sages down through the ages" have uncovered concerning our investigation. Part Three then wrestles with modern developments in this area of research. In this way, the study follows the threefold division of my prior work. Yet this is where the similarity ends, for three additional parts round out the treatment and the arrangement is thereby doubled. Part Four considers more current analyses in the field, Part Five reviews two documentary films as case studies of what humans engage in and what this says about us, and following the Conclusion where I interact with much of the foregoing material, the Appendix amounts to a recounting of two dreams from yours truly as a further indication of the type of experiences that are common to humanity.

Such a format, it must be acknowledged, might not readily lend itself to continuous transitions. This is the risk one runs when themes are many and varied, for one will not necessarily lead or flow smoothly into the next. This is not an apology but an admission as well as one rationale why this marks my first preamble with my current publisher before otherwise proceeding straight into an introduction. This allows me to say what I believe really needs to be stated, where the text to follow might not readily facilitate it.

These matters are of note in our endeavor: I say the exercise is ours because I invite the reader to follow along and perhaps press on as problems are tackled that have no definite present or maybe even projected resolution; the title is taken unabashedly from the 1973 record album by the rock group Deep Purple, despite their not having had the same concerns as mine; and the presentation often exhibits a decidedly negativistic posture on human nature, thus critics insisting on equal proportionality will be disappointed. On a personal note, I am afraid that I cannot be entirely positive about the prospects of human nature. We have all come to know what humans are capable of, both the good and bad, and I wager whether the default drive of the bulk of humans, particularly in times of crisis (though our best characteristics have been known to shine forth here as well) is geared toward the negative end of the scale. It is a bad sign when the media, which does not fail to report on and present the negative side of life (and what does this say about those who feast on it?), points out acts of kindness as newsworthy owing to their scarcity. Bus drivers and police officers who procure shoes for the homeless and marathon runners who quite literally go the extra mile, or in this case two, to a hospital in order to give blood for victims of bomb blasts are rightly lauded as model citizens, when their example should be more the norm. But more often than not, at least this is the way it seems, people will act out of self-interest and some out of hate. For reasons such as these, I see the pessimistic view as the camp where greater accuracy resides. Rebuttals are invited.

On to another form of division. There are those who understand humans as composites, made up of multiple components. Many in the Christian West still interpret humans as a threefold organism, composed of body, mind (or soul) and spirit, conferring on the creature the term tripartite. Others condense the latter two and proclaim that humans are dipartite, made up of two things, physicality and mentality. For those in the second camp, everything extending beyond the physical is apparently relegated to the same category, no matter how diverse it or they might be. Others still see no requirement for holding that humans are compounds, but are entirely physical (the materialist position) or mental (idealist). Each of these alternatives will be scrutinized.

A word of caution as we launch into the main work in earnest. The introduction that follows directly is not of the type we have come to expect, hence the need for this preface. What I introduce comes in four segments, all of which will have a bearing on the topics to come. I ask the reader's indulgence for the seemingly episodic nature of what follows. In this way, the body of the text reflects the bodies and persons we have become acquainted with, though this acquaintance by no means indicates a familiarity with the intricacies of the subject matter. For there is slow progress in coming to terms with human nature, and belief that we have a lock on it is hubris. We might assume that we, being us, are experts on us, but that would be presumptuous. This is why we need an extensive look as to who it is we think we are.

Introductory Material:
Four Statements on Human Nature

The first installment in what might best be described as a series of introductions or vignettes, all geared to give the reader a taste of what lies ahead, is this.

My heritage is German, Germany being one of the two nations to have produced the most renowned philosophers in the history of the discipline. The other is Greece, though their philosophical heyday has long since passed. Ordinarily, I possess the trait of systematic analysis, which, I imagine, is a legacy that my heritage has bequeathed to me. Yet I also have the dubious distinction, perhaps contrary to this same legacy, of not being able to figure my way out of a jam when it comes to certain electronic equipment, namely remote controls. It is not so much that they describe a foreign language to me as that the sequence of events needed to be applied so as to carry out an intended task does not in my mind resemble how one in moments of clarity should go about it. The geniuses who decided that this should be the way to accomplish a feat, such as installing a program, are far-removed strategically from how I would have done it. My mentality simply cannot fathom theirs.

Thank goodness for my wife. She is of Dutch descent and thus unencumbered by an unalloyed Teutonic mindset, though the genetic markers are not so different. Yet she is able to accomplish what I cannot. Hers is the more intuitive of our natures. She can wade her way through the mind maze intuitively, an outrage to one of a more logical bent. When faced with a programming quagmire, she steps in and I step aside and find another project to concentrate on. She can unlock, disentangle and unravel the mystery. We each have our strengths and weaknesses and hence contribute to the team effort.

Why do some people have this talent and others that? And why are there as many mindsets and perspectives as there are persons, especially when we feel strongly that the way we see things is the obvious one and ought to be universalized? The complexities of personhood defy straightforward analysis. Different people react differently to similar situations. Thinkers throughout the centuries have grappled with what makes us who we are, with only partial success. Their views might be applicable in part, but they never quite capture the whole. No one approach offers a complete picture.

We may have different characteristics but the same human nature. For this reason, we can benefit from the talents of others who can complement our own, and where their efforts can compensate for our shortcomings. Part Two will highlight historical attempts to grasp the nature of humanity. Part Three will outline some contemporary proposals. In the meantime, I should consult with my wife about the finer points of convection ovens.

Here is an example of human nature in action. At the College where I have been employed on most occasions, there are the familiar mail boxes in the faculty lounge, where communication can be made with faculty through what appears to be never outdated paper means. Students often submit their written work in these slots when unable, or unwilling, to appear live and in person in class. What deviates from custom, or perhaps habit, is that deposits must be made below the stated name on the pigeon-hole and not above it. This leads to periodic difficulties, for a sizeable proportion of submissions is placed above the name, thereby prompting me to ask if there is a natural inclination to place items above names?

Upon reflection, the reasoning employed on the part of those following the dictates of these proposed natural inclinations might run something like this: the edges of the pieces of wood separating slots is where names are placed, hence the piece of wood upon which a name appears is that belonging to the faculty member in question. The name indicates at least temporary ownership. Thus to submit an item on top of the piece of wood is to place it in the care of the name on which it sits. And by the same token, to place an item below a name would be to place it on someone else's territory. That would be an infringement, a failure to recognize the personal property of another.

Whether or not this is the case, there does seem to be a tendency to follow the lead of others. When one ignores convention and places written work below the name, others may do the same, regardless if any do so with conscious intent. None might reflect on the matter, but if they do, they may opt for the "safety in numbers" policy. Choosing this alternative means all are in the same boat and would experience the same fate. There is comfort in this, cold though it may be.

Which of the above two possibilities trumps the other, or do they work in tandem? Are people more likely to follow the leader or have a high regard for perceived personal property? Issues of this sort are touched upon below.

We seem to have another tendency: we are given to evaluate relationships based on the last encounter that was had. This becomes a type of "what have you done for me lately" approach. If there are those we have not seen for a lengthy period, our impression of them can be our most recent or most memorable interaction with them. Should that encounter have been positive, then we might be inclined to think, "Golly, what a swell guy" (at least we might put it this way if we were stuck in the 1950s). Alternatively, if our latest interaction were to have been negative, then we may be moved to bemoan, "Well, s/he was always a miserable sort anyway." Our perception then proceeds in reverse chronological order, namely the distant past is appraised in the same light as the immediate past.

A person's life then, fairly or unfairly, is defined by his or her most recent or most publicized act. Pete Rose will forever be remembered not for his hall of fame caliber baseball statistics, but as the high profile former player who wagered on the sport while he was still involved in it (he bet on games while in a managerial capacity). Certain disgraced politicians' careers are also judged by their most recent errors in judgment while in office.

Nor does it end there, for we assume that their disposition extends from the womb to the tomb. Our evaluation, as already intimated, becomes retroactive. Our initial alarm is replaced with an analysis of why a life such as this could have proceeded in no other way (in history this is known as the historicist approach). Assessments such as these can become entrenched and thereby difficult to dislodge.

I suspect that a similar thing might have occurred when it comes to the honor accorded to the one alleged by Christians as Messiah. The final encounter, or period of time, that the disciples had with Jesus, according to the accounts, were the post-Easter resurrection appearances. This colored everything they thought about him. If there would have been no perceived resurrection events of the type recorded, then his life, for many of his followers, then and now, would have been interpreted as an abject failure, a nice guy who emphasized righteousness and God's kingdom and got killed for his trouble. When he announced that the way of the religiously self-righteous was not God's way and that the Temple establishment they depended on was idolatrous (both themes that prophets before him stressed), he was a marked man. The Roman objection was that Jesus set up God's kingdom as a rival authority to Caesar's. The emperor would never countenance being a subordinate and Jesus was executed for his insubordination.

Yet it was not entirely like this. His disciples interpreted the resurrection appearances as legitimating all that went before. He was right all along, the sentiment became, and God validated his mission. Everything that came before must then be viewed in a similar light, namely the presence and power of God must have been manifest throughout his career. His words and actions must then be seen in a transcendent way, extending to his entire ministry—perhaps even for events that never actually took place as well as for what might have occurred prior to his public appearance. Their reasoning could have prompted the following discussion: "That's the kind of thing he would have done or said given half a chance, so let's put that in his story too. No doubt he was a precocious youth, so let's have him take on the rabbis at the temple at Bar Mitzvah age. And what would a Messiah be without a miraculous birth? Other religious figures have had them, so why not him? And if he calls God 'Daddy,' then he must have had a heavenly beginning, even before birth. He may not have said these things in so many words, but we will do so on his behalf. What better way to honor him?"

So the resurrection, for them, could have entailed all these things. "No Messiah," they might add, "would rest on the resurrection as a stand-alone event. Much of note must have led up to it. The word on the street is that he did many anointed-type things before his death. So let's find them. Leave it to us; we will tell it the way it needs to be told." And so on.

This is how an exemplary life becomes a product and how a corporation gets built up around one. We are left to wonder if Jesus would have recognized himself in the accounts drafted about him. Would his biography match his CV? In our exuberance, we tend to embellish our heroes and our recollections reflect the larger-than-life.

In the opposite vein, if there is an episode in someone's life whom we hold in high regard that we would just as soon forget, the report could resemble a literary sweeping-under-the-rug. A case in point concerns the fate of Josiah, king of Judah. He was heralded as a messianic-style figure who instituted extensive reforms, all in an effort to bring the people back in line with God's commands. Being hailed as a new David, perhaps he came to believe his own press clippings, for in an ill-advised manner he went out to meet Pharaoh Neco II of Egypt, who was marching out to assist the king of Assyria. The record of their encounter is this: "King Josiah marched out to meet him in battle, but Neco faced him and killed him at Meggido"(2 Kings 23: 29). Short and definitely not sweet. Admittedly, the account in 2 Chronicles 35: 20–24 is somewhat lengthier but also deviates from the former, in that it has Josiah only wounded from the battle but dying in Jerusalem. That would constitute a very slow killing on Neco's part.

Victors tend to be more lucid and to glory in their successful campaigns. They have more to say. The vanquished, on the other hand, would rather not recall how they were conquered. They are more reserved and laconic. They adopt a policy of "the less said about it the better." Can you blame them?

Reporting is always from a certain perspective; there are no purely objective independent observers, and our vision is always distorted by the lenses we use to experience it. One person being interviewed was heard having said, "I just get the straight news, I watch FOX." Regrettably, neither this brand of news nor any other can offer the straight goods, since we all have our biases and prejudices (with FOX and CNN perhaps having a disproportionately larger share). All forms of communication are agenda-driven, including this one. Not to worry, though, for this is inevitable; it reflects the way we operate.

We are informed, even cautioned, that "If you can't do the time, then don't do the crime." Words to live by. Best to stay out of prison at all costs. But it does not work the same way with pain. Some pain is unavoidable, no matter how well we live. No one appears to be immune to it, for to live is to experience some form of pain. And some of it is undeserved. A patron of a Montreal café sits outside and is felled by a falling slab of concrete that became dislodged from a building. Regardless of the moral stature of the person, no one deserves this. Such an end comes not from living a less than stellar life, but from living in a complex and messy world, where gravity can trump safety or convenience, irrespective of who or what lies beneath. Sometimes we are in the way. The results can be tragic.

A healthy person might be told, "You must live right," as though there were a one-to-one correspondence between living a life beyond reproach and reaping benefits or rewards from it. Contrary to this notion, however, lifestyle does not determine personal outcome. My father was diagnosed as having a cancerous growth in his colon, for which a bowel resection was ordered. While the procedure was successful, the cancer had already metastasized to other less readily treatable organs, and this became a factor in ultimately claiming his life. This now places me in a higher risk category, since I have a family history of the disease. So off I went to endoscopy for a colonoscopy of my own. The same physician who treated my father also saw me. Like father, like son in some cases but not in others. The news was good—no signs of cancerous growth.

The upshot of it all is that my health is not always dictated by my choices, for some of it is a gift from my forebears, though they need not have bothered. While I was awaiting the results, the anticipation afforded me ample opportunity to reflect on better days, in the hope that I would still see plenty of them. Odd how some things seem to occupy one's thoughts and fill up all the available mental space. I admit that I had difficulty focusing on anything else at the time. We have heard it said that "we need to pay for our sins somehow." We can only hope that any undeserved suffering we experience counts as "time served."

The foregoing are intimations of what can be expected in the pages ahead.

Part One

Biblical Material

REGARDLESS OF THE PERSPECTIVE readers of the Bible bring to the text, what is clear is that the Bible makes claims about God, humans, the rest of the material order, and the interrelation among all three. Of interest in this study are the claims made about human nature. Part One will address the theme from a literary critical standpoint and Part Three will treat the topic in a more philosophical vein. Biblical statements here will be examined in their order of appearance in the Judeo-Christian scriptures. Quotations are from the New Revised Standard Version (NRSV) unless otherwise noted. (Also, any italics embedded in quotations throughout these pages are those of the authors cited unless otherwise specified.) Parenthetically, this marks the first time that I have not employed the term God in the title of my books, though the reader can expect that it will figure prominently at many a turn.

Each Hebrew theological category had a beginning, from the pen (or a stylus of some description) of a Jewish author or authors. There was a time prior to which the Jews had a creation story. They first had a notion about their own history and how they became the nation of Israel. The trouble is that the account of their having relocated from Canaan to Egypt, their mistreatment under a certain Pharaoh, their exodus from Egypt and eventually their sojourn back to Canaan after forty years of wilderness wandering is unsupported by the archaeological evidence. The Hebrews might actually never have left Canaan as such but merely separated themselves from already existing Canaanite cultures and nurtured one of their own. They may also have amalgamated with

2 Who Do We Think We Are?

a group that itself migrated from Egypt and incorporated aspects of both of their histories into a single combined chronicle. The upshot is that the bulk of the Jews were possibly fellow Canaanites estranging themselves from other Canaanites (Finkelstein & Silberman, 2002 and 2006). (As always, these views are not without their detractors.)

Having established their national identity, the next step was to decide how humans and other nations arose. Then finally, to complete the picture, they recognized the need for an account of how the heavens and the earth got their start. Thus with the passage of time, the Jews commented on what occurred in their past, an ever more remote past in each subsequent case. First emerged a socio-political report of their own people, the children of Israel, second an anthropological reading of humankind in general, and lastly a cosmological rendition of the world itself. The national creation story crystallized likely at the time of king Josiah in the seventh century BCE; the anthropological creation legend probably during the Babylonian captivity in the sixth century BCE; and the cosmological creation myth shortly thereafter. Spurred on by competing traditions from the surrounding cultures near which they found themselves, particularly threatening while in exile, they responded in survival mode (anticipating Darwin) and proposed their own versions. Both the people and the stories endured.

Hence theologies evolve. It took time for these ideas to simmer and become committed to writing. Before written works could emerge, at issue was the elapsed time Hebrew history underwent for their language to move from merely oral tradition to script and then for literacy to abound. As one example of theological evolution, a belief in an afterlife beyond the gloomy concept of Sheol, which had been standard fare for some time, and where the shades of individuals simply corroded away, was not accepted until the Maccabean revolt period of circa 168–142 BCE. During this time, the second half of the book of Daniel, perhaps the latest writing contained in the Hebrew canon, was completed. (By way of explanation, some suggest that the book of Esther is even more recent, and a canon is considered to be an authoritative body of texts.)

The perceived need on the part of the Jewish people at the time was an understanding as to how the ill-treatment of the Jews by the ruling Greeks could be reconciled with a belief in God's justice. The resolution was a notion of a resurrection of the dead back to life, whereupon God would settle accounts and exact revenge on those deserving of it in a life beyond this one. God will set things right in the hereafter (Daniel 12:2), when God's justice will be made entirely manifest. But this doctrine took a long time in developing. Were there no shortcuts?

If this is such a crucial doctrine, the question must be asked, then why did it take so long for it to be introduced? Did God work with God's people only on a need-to-know basis? Or was there difficulty in the transmission such that God could not effectively get God's point across? Were the people slow to grasp, or did they make their own unaided decisions? The history of Christianity reveals that humans do the deciding, sometimes after long periods of debate, as to what constitutes orthodoxy. The point is that this state of affairs, one hesitates to say "truth" when it is this fluid, was not so much discovered or received through revelation as it was affirmed. Humans are a factor in the proceedings, perhaps even the main one. Despite divine input, "For it seemed good to the Holy Spirit *and to us*" (Acts 15:28) (italics mine), humans are a variable in the shape doctrines eventually take. The forms which both documents and doctrines assume have multiple causes, and there is much uncertainty along the way. And this uncertainty remains since human authors never fully capture the truth. This is where we find ourselves, which is why there is a call for theologians with every passing generation. The need for humans to refine the ideas of other humans never stops. Part of the human condition is to "see in a mirror, dimly" (1 Corinthians 13:12).

Joshua and the conquest of Canaan: Joshua 3:10—"the living God who without fail will drive out from before you" all the inhabitants of the land.

What I am about to say is likely the most controversial statement I have ever made in print. The following subject matter contains violence. Reader discretion is advised.

4 *Who Do We Think We Are?*

There is no archaeological evidence that the children of Israel displaced the people of Canaan and occupied the land in their stead. Rather, the Hebrews seemed to have been Canaanites already and at some point decided to remove themselves from the lowland populations and become separate by retreating to the interior highlands. (See the aforementioned works by Finkelstein and Silberman). This is prefatory to my point which I declare with much trepidation. But what follows is the kind of thing humans have been known to contemplate and execute. I broached this topic in my previous volume and its impact will not leave me alone. It stays with me like a protracted case of heartburn. Maybe the analogy is warranted. Perhaps in relating my disgust I will find relief.

I begin with autobiographical background. Both my parents (now deceased) were German. This is my heritage. My mother lost her first husband in the Second World War, and her daughter, an only child, was my half-sister who died thirteen years prior to my birth. She perished in war-torn Germany where medical attention was denied to those families refusing to join the ruling party.

There is much in the way of music and philosophy that Germany in its history has bequeathed to the world together with many additional scientific and technological advancements in more recent times. Sparking a second world war was not one of them. What certain Germans perpetrated in this theater is a blight on their cultural landscape, a black stain on a white sheet on which the eye invariably focuses. I am not proud of this part of my heritage, especially since it is such an overwhelmingly powerful aspect of German history that it seems to cloud the rest, no matter how admirable that might be. The atrocities committed by the regime remains a stench in the nostrils of those adversely affected by it. The ones who survived to tell the story rightly remind us of the evils concocted so that they will not be repeated. And if I am not careful, I could fall into the trap of becoming ashamed of all things German.

A negative event can psychologically erase much good. This is a human tendency already alluded to. Adolf Hitler was a racist who despised the Jewish people and sought to exterminate them, thereby catapulting him into the ranks of the most hated person in Western history, though Josef Stalin could also vie for top spot. This can prompt those reflecting on the events to ask, "Can anything good come out of Germany?" Yet, and this is the controversial part, is the attempted extermination of peoples always a source of shame?

Turn the clock back almost three and a half millennia when the Hebrews allegedly escaped from Egypt and eventually re-entered the land of Canaan. The Bible paints a certain picture of how the events unfolded, yet they are a retrospective and therefore likely suffer from a condition known as "spin." If the occurrences were historical, a running commentary might proceed as follows. The God of Israel commissions the people to enter the land of Canaan, for it is to be their inheritance. Now comes the unsavory part. Never mind being given the land, God commands them to take it. Joshua then embarks upon a series of campaigns designed to oust the inhabitants by force. In fact, God's chosen people are to exterminate not only the inhabitants but their livestock and everything that breathes; they were to be shown no mercy (as if to say that cattle should have been more circumspect in leaving their life of sin) (Deuteronomy 7:2; 13:15; 20:16–18). Hence extermination by the Hebrews is embedded in their very narrative tradition. In these accounts, the children of Israel are the aggressors.

We can only hope that the priesthood which later reflected on these events and committed them to written form for posterity embellished the reports as victors often do. That the extermination was carried out is something we pray is hyperbolic. At this point we cannot determine whether the God of the text is vindictive, or the people misinterpreted the directive and overreacted. Could not the procedure have begun with some diplomacy and the offer to the inhabitants to "join us"? And only if they refused or attempted to coax the Israelites to adopt their idolatrous ways would the condition of "or else" be added.

This brings us to the question of the historical accuracy of the accounts. If certain archaeologists are correct that there was neither an Exodus from Egypt nor a conquest of Canaan, then the drafters of documents such as these have some explaining to do.

If they did not occur but the authors were eager to include them in the text regardless, then they are in danger of being charged with misrepresenting God. How then could they be trusted to act as spokespersons for God in any other matter?

There are three suspects here: God, if God operates according to a different standard and maintains that one should refrain from killing unless God orders it; the people, if they took the issuing of God's command as a license to "take out the cultural trash"; or the priesthood of authors, if it elected to give false witness/testimony about God. My inclination lies with the frailty of human nature, meaning the second or third alternatives are the likely candidates.

I began by lamenting the atrocities of my heritage in attempting to exterminate a people for reasons of hatred. Here, entertaining for the moment the possibility that the accounts reflect actual historical events, we are driven to the realization that the race of persons who endured their near-extermination also formerly inflicted it, though perhaps not for reasons of hate. The same race would then both be on the giving and receiving ends. If Joshua's strategy arose from humans, then it would stem from hatred; should the directive have come from God, however, then God's justice borders on ethnic cleansing. Neither option seems appealing.

The Jewish people are rightly proud of their heritage, but Joshua's bloody campaigns would then be part of it. Those who objected to ill-treatment more recently are the same lineage as those who exacted it more remotely, though there appears to be no shame attached if the perception is that they were acting on God's behalf. Besides, if Joshua's exploits are not actually part of their history, then why invest in it? In an age of purported tolerance, perhaps it would be better to distance oneself from it.

That's no way to treat a woman (or child for that matter): Ezra 10

Early in the sixth century BCE, the Babylonians "came, saw and conquered" Jerusalem and carried off captives in a series of three deportations, starting with the leading figures of the city and ending with the ordinary, thereby leaving only the poorest peasants in the land. Those Jews who later reflected on this tragic chapter in their history, tragic for they were driven from their homeland and the temple as their central place of worship, saw this as a retributive act on God's part. God, through the prophets, had warned the people repeatedly to reform their ways, stop worshiping idols and serve the one true divinity. But having turned their backs, God demolished their greatest source of confidence, said temple, and drove them far away from the land that had become polluted by their disobedient acts. These chroniclers interpreted the events as having been directed by God, who uses nations to exact punishment on other nations. "Babylon vanquishes Jerusalem," the headline read.

About a half century later, when the Babylonians themselves were succeeded by the Medes and the Persians, Cyrus, king of the Persians, in an outbreak of tolerance decided to release the captives and allow them to return to what was left of their homeland. For this policy, apparently customary on his part, he was viewed as messianic. Some took him up on it and went back to rebuild; others remained behind since they had already built a good life for themselves there in downtown Mesopotamia. Later generations also made the trip back.

It appears that the contingents who returned comprised a more conservative element. Upon their return, they rebuffed the overtures on the part of the Samaritans who offered to assist in the rebuilding project. The Jews regarded them as enemies, uncharitably perhaps, and claimed that they have no part in Israel (Ezra 4:1–3). Almost two centuries prior to this, in the late eighth century, the northern kingdom of Israel had been taken into exile by the Assyrians, and those who remained were compelled to intermarry with their captors. In the eyes of pure Jews, this irreparably diluted the bloodline and forever created a rift between Jew and Samaritan, hence the scandalous nature of Jesus' parable of the Good Samaritan. Long live the grudge.

Then Ezra comes to town. He was a legal scholar dispatched by the Persian king at the time to travel to Jerusalem in order to sort out some difficulties there. He along with some other Israelites, including priests, Levites and temple officials, were sent by the king to rectify the ignorance about, or non-adherence to, the law of God and the king on the part of the people. Offenders were punishable by the magistrates whom Ezra was to appoint (7:25–26).

It soon came to light that the people had fraternized in a nuptial way with the surrounding cultures. The "holy seed" had become mixed (NRSV) and the "holy race" became mingled (NIV) with foreigners from whom Israel was to remain separate (9:1–2). One wonders if this policy of racial purity came from God or was the position of the conservative majority. I say majority because the sentiment was not unanimous, though only a meager four constituted dissenting voices (10:15). The decision was reached that the foreign women as well as their offspring, being but half-Jewish, would suffer banishment from Israel. The authority for it was the king's command (7:26).

Such was the fate of those not following God's or the king's law. This assumes, however, that the foreign women had not given up their "detestable practices" (9:1) in favor of the perceived, or then understood, ways of God. And even if they did not, what was their crime? Any wrongdoing was on the part of those who took them in, yet the women and children ended up suffering the reprisals. A possible further injustice could occur if the women would not be welcomed back to their native land. Such a move might also not bode well for the offspring, for, since half-Jewish, they may not find a home there either.

Was ignoring compassion a faithful administration of God's law, or a perverse sense of devotion to the letter and not the spirit of the law? Was God pleased with the outcome, or was this only conservative fanaticism? Legal rigidity seems to yield further injustice.

Divine wagers: Job

The appeal of making a wager lies in the excitement of prognosticating the future, together with the spoils one can expect to receive upon guessing correctly. If you have confidence that you can "divine" what the outcome of an event will be, you might be prompted to put your money where your mouth is and place a bet. Or even if you lack such confidence, the thrill alone can be sufficient to make the experience worthwhile.

Based on the foregoing, what are we to make of the book of Job? We meet here for the first time in the biblical accounts, at least in the sequence in which the texts appear in the table of contents, the figure of the adversary, or the accuser, commonly known as the devil, Lucifer, or Satan. The devil does not seem to be out of place here, but is an accepted part of God's entourage. As a type of prosecuting attorney, the devil, whether on his own or commissioned to do so—we are not informed—opposes those whom God favors. The devil challenges God to a bet concerning the loyalty of God's servant Job. Satan provokes God by stating that God's pride in Job is unwarranted, for simply by taking away Job's possessions, including some family members (Satan thereby banking on typical human nature), Job would reveal his true allegiance, one which is not geared toward God. God takes him up on it, Job loses his livestock to theft, and his offspring and servants to death, yet remains faithful to God.

God wins the bet, but Satan is undaunted. The devil declares "double or nothing" and maintains that if Job's own person were to be afflicted, for anyone's most precious possession is one's life, he would renounce God. Jeopardize his life and the offense will make an enemy out of Job, Satan bellows. God says, "you have a deal, 'only spare his life'" (Job 2:6). Once again, Job does not falter, the devil is disappointed for a second time, and God wins both bets.

Three things to mention. First, usually in a wager, one of two parties stands to benefit. Where in this case is God's take? Admittedly, God obtains the satisfaction of having a refined all-star super disciple, yet otherwise only Satan stood to gain and God to lose, for Job was already touted by God as the world's most righteous person. God should have contested that Satan ante-up with something of equal value of his own so as to make this a two-way bet. The moral of the story, I suppose, is that there is nothing that compares to a faithful servant. Second, the story seems to imply that God was unaware of what the future would hold. This casts doubt on God as possessing foreknowledge of events, otherwise it would not have been a fair bet. If God already knew the outcome, then the God of the text is about as deceptive as Satan, who is known as the deceiver. Third, Satan should be accustomed to God's capabilities by now. He should recognize God as having certain powers. Satan may be evil but s/he is not stupid; s/he would not enter a wager situation if the outcome were to be a foregone conclusion, for that would not make for a level playing field.

So it appears that a wager took place where neither participant foreknew the result, the only true type of bet. Both parties pinned their hopes on human nature: the devil believing that human frailty would win him the bet; and God that this particular human's faithfulness would shine through. God is revealed as the true judge of character. Yet these two main players take on the roles normally associated with the Greek and Roman pantheon of deities, replete with human flaws. God in this case being taken in by the devil's wiles, accepted the challenge and agreed to play the game, though with conditions, on Satan's terms. Does this instil confidence in the managerial capacities of the spirit world?

Or allow me to approach the story from different angle. In the exchange between God and the devil, God asks Satan if he has noticed God's poster boy, the world's then current most exemplary human (Job 1:7–8). This is the type of banter that a couple of guys could have over a beverage. And in the event that these libations have contributed to memory loss, a similar conversation between the two interlocutors is repeated in the next chapter. Now of course the divinity is held to boast total recall and no amount of fermentation sampling can alter that. More than this, God is also alleged to know both the course and outcome of events, or at least be in the best possible position to anticipate them. Nevertheless, there evidently is some uncertainty in future results that cannot be ruled out. Hence the drama. The author must be alluding to the notion that certainty is unattainable, even at this level, otherwise there would be no story. The suspense must be real for all parties concerned, else why bother going through the exercise? As mentioned, Satan, no doubt having sized up his opponent, would have known that entering contests of this sort would be futile on his part.

Despite this being a work of fiction, there are those, as we will see in Part Two, who believe that destiny does play a role in human lives. But if so, then we must ask, with what kind of regularity? If this type of situation occurs numerous times in history and is not confined to the book of Job, then either Satan doesn't get it, or God has a gambling issue, or both. Either way, the divinities are made to look like the Greek and Roman gods—all too human.

Gospel interlude

Skipping well ahead, a few words about the gospels in general before we continue. To begin with, there were several of them, only four of which were accepted into the canon. This means that there were others around at the time not in the global ecclesiastical fold, when the four successful ones were circulating. Apparently, this situation might even have become a spur for some individuals or groups to prepare their own. Perhaps these persons were sufficiently critical of previous attempts at gospel writing, "Since many have undertaken to set down an orderly account of the events that have been fulfilled among us,..." (Luke 1:1), they were prompted to draft another which would contain those emphases most pertinent for their intended audience.

Next, sometimes there exists a tendency on the part of gospel writers to be free with the words of Jesus; some might even be compelled to say "taking liberties." The strategy may have been to present what Jesus would have said given the opportunity, in essence putting words in his mouth. The authors may have understood themselves as fitting squarely in to the Jesus tradition, but by modern standards this is looked upon as embellishment or outright misrepresentation. Even the Ten Commandments warn against bearing false witness (Exodus 20:16). Or perhaps they reasoned that this did not legally apply to them in this circumstance. Besides, they had God as their role model, for the God of the text even transgressed this directive. On one occasion, a prophet declared that God had sent a spirit as a lying influence upon a king's own company of prophets (1 Kings 22:20–23). So this God is not above using deception as a means to get the point across. With a textual authority like this, the gospel writers might have seen fit to craft the type of account that would best fulfil the purpose of their mission, namely to submit an appeal to come along and follow this Jesus too. In this sense, they had an agenda to attract others to Jesus, and sometimes this agenda trumped historical accuracy. In any case, it must be kept in mind that there is no requirement on the part of these authors to conform to modern standards of scholarship, for that is a convention which the contemporary Western world has imposed upon them. They cannot be faulted if in their context there was not the same obligation.

Turning to the Jesus of the gospels, who he was, what he did and what happened to him are not just modern debates, for they were argued even in his own time. His wisdom and the wonders he performed were too much, in the eyes of some, to be attributed to an ordinary carpenter's son. Thus, some became offended at him. Consequently, this lack of faith diminished the amount of effective work he could accomplish in a given setting (Matthew 13:53–58). The Pharisees even concluded that his abilities were demonic in origin (Matthew 9:33–34). The debates have raged on ever since, largely because of a realization as to what is at stake. If the gospel writers are correct in their assessment of him, then he may arguably have a claim on our lives. This is a topic for another discussion, but suffice it to say that issues surrounding Jesus' life and identity are not new.

There is at least one instance in which I have difficulty with the Jesus Seminar, a group dedicated to tracing the initial instance of the recording of Jesus' words so as to obtain a clearer picture of him. They generally believe that the first document to do so is what is dubbed the Q source, Q being the first letter in the German term Quelle, meaning source. Evidence for the existence of this non-extant text is held to be drawn from the gospels of Matthew and Luke. There are episodes in these two gospels where they probably used the gospel of Mark as a source, since they agree in the material that they glean from it. Yet there are also spots where Matthew and Luke agree without any borrowing from Mark, suggesting that there was an additional source to Mark that both of them employed. Sounds likely, but it is not automatic. My misgiving about the approach of the Jesus Seminar is, in my view, the misguided assumption with which they operate. They assume that the closer one gets in time to an event, in this case the very words of Jesus, the more accurate the portrayal drawn of him. This is what Q, for them, achieves. As we noted above, though, the debates about Jesus even at his time do not clarify matters for being contemporary with him. Not everyone held the same view of him even then; indeed his observers and commentators were deeply divided. The Jesus Seminar, then, fails to appreciate that, in words we have already used above, even the Q source depiction of Jesus can be as agenda-driven as any other.

Lastly, the earliest writings which later became incorporated into the New Testament were those not of the gospels but of the letters of Paul. Scholars such as L. Michael White (2010, 123) have wondered why this apostle would omit vital information about the birth, life and details of the death of Jesus, a task left to the later gospel writers. Is this not a case of putting the cart before the horse? First things first, these critics might say. And while further details, or at minimum confirmatory ones to those in the gospels, would be helpful, I submit the following remark. I find that Paul is not required to recount first principles of Jesus' ministry, for his writings could be building on an already assumed oral foundation, sparse though it may have been. What Jesus did and taught might have been regarded shortly after his death as sufficiently well-known and understood, such that at least for Paul it needed neither introduction nor retelling. As proximity to these events receded, however, and as the message became broadcasted more widely in the Roman world where it had not already reached, the need would undoubtedly have arisen. There would then have been warrant for such a treatment. Hence the arrival of the gospel tradition.

Agreeing to disagree:

Now for the heart of the matter. In the role of college instructor, I was challenged on one occasion by some students, while leading a seminar in a course on the philosophy of religion, to present examples of instances in which the Bible exhibits internal inconsistencies, the assumption on their part being that it does not. Defiantly, these students proposed that the Judeo-Christian scriptures are happily free of contradiction. Regrettably, I became non-plused and drew a blank when called upon to counter their claim, bewildered that those holding such a view had not as yet gone extinct. The scriptures which they claim as their authority to the point of idolatry (the affliction known as bibliolatry) speak to this malady, though in a different context, and it can be applied here. Paul in Romans 10:2 laments that his religious conspecifics, the Jewish people, "have a zeal for God, but it is not enlightened." Precisely. We should seek to disabuse these students of their error.

As seems to be typical of human nature, in the intervening time several discrepancies have resurfaced for me. They are not very difficult to find, but these students have long since departed for the glories of the world beyond academe. A missed opportunity on my part. Nevertheless, in the following I submit my long overdue assignment. Since it is my policy always to have students hand in something rather than nothing at all, and better late than never, I now go public with the exercise. This is for those students and everyone else who can benefit from it. The list is not intended to be exhaustive but representative and instructive, sufficient to make the point.

Ask people what happened at the scene of an accident and you will probably get as many perspectives as there are witnesses giving them. There will be discrepancies in reports, as courts of law know very well. Since humans do not come with infallible sight or observation skills, it should not surprise us that inconsistencies crop up, if not abound, not only in official records but also in the Christian sacred text.

As an initial installment, let's focus on the issue of the visibility of the divinity. Doctrinally, spirits are supposed to be invisible, for 1 Timothy 6:16 emphatically declares that God cannot be seen, so no one has. But is that the end of the story? Astonishingly, the most blatant example of inconsistency occurs not only in one biblical book but in the same chapter within it. As a prelude, Exodus 24:9–11 announces that Moses, his brother Aaron and seventy-two others ascended Mount Sinai in order to meet with the God that entered into a covenant, or contract, with the people and had given them the Law, including the Ten Commandments. The episode reveals that God in fact was *beheld* by this contingent, who "ate and drank" or had a picnic there. Not a common occurrence, perhaps, but God allowed it to take place. Then in chapter thirty-three comes the concern. We are told in verse eleven that the interaction of God and Moses was on a friendship level, where the two met "face to face." Verses 19–23, however, tell a different tale. There it states that God's face cannot be seen, else the observer dies. God's face here is off-limits. But God's face has already been seen. Leaving little room for doubt, Numbers 12:8 concurs that Moses saw both God's face and form each time the two of them conferred at the Tent of Meeting. So which is it? If God and Moses already had a facial encounter, then why claim in the same breath that this must not happen? Isaiah even affirms that it is possible to see God and live (Isaiah 6:1–5).

For those who suspect that God at this point simply proclaimed that enough is enough and that the appearances shall cease, there are additional passages to consider. The episode about Jacob, who also made eye contact with God (Genesis 32:30), is an alleged event that occurred much prior to Moses, so the case for the defense is still intact. (Moreover, Deuteronomy 34:10 is simply a restatement of Exodus 33:11.) Ah, intones the prosecution (that's me), but turn to the end of the book of Job. There God speaks to Job out of the whirlwind and gives him an earful about who is in charge. In his response to God, and here is the crucial point, Job states that he has laid eyes on God: "I had heard of you…, but now my eye sees you"(42:5). Taking a passage from a different biblical book, the psalmist rhapsodizes that he has "looked upon [God] in the sanctuary"(Psalm 63:2), and if seeing God in a vision counts as evidence, then Daniel can claim this as well (Daniel 7:9).

There is another episode which may or may not fit into this category. In Genesis 18, Abraham receives three male visitors and instructs his wife Sarah to prepare a meal for them. One of the men is often referred to by the author as the LORD. When informed that the visitors have come to destroy the cities of Sodom and Gomorrah, if the rumored wickedness in them is verified, Abraham pleads their case. He negotiates with the LORD to spare those places for the sake of the righteous in them, even if they number only ten. Abraham appeals to the LORD's sense of justice by asking, "Will not the Judge of all the earth do right?"(verse 25 (NIV)) This strategy proved successful. The trouble was that not even ten righteous persons could be found. Abraham should have asked for even fewer. The point is that if this Judge can be identified with God, if more than simply an angel, then Abraham most assuredly had a visual of God and spoke with him live and in person. Perhaps this also fuels the tradition of God as male.

Despite these instances, three New Testament authors insist that while there are audio portions to divine revelation, there are no video. John's gospel flatly denies that anyone has seen God (1:18); to reiterate, the writer of the first letter to Timothy (probably not Paul) contends not only that God has not been seen but cannot be seen (6:16); and 1 John 4:12 rounds out the triad in similar terms. Interesting that the battle lines are drawn roughly between the Old and New Testaments, with the deity's visibility diminishing with the passage of time. The change in emphasis might depend on the doctrine that the writers intended to defend. It appears to be human nature to shape information for the aims one has in view, in ways that will bolster one's own position. But wait, there's more to be said and on a variety of themes.

For another discrepancy, all three Synoptic gospels (Matthew, Mark and Luke) agree that the figure of Simon of Cyrene was compelled to carry the cross as Jesus was on the way to his crucifixion (Matthew 27:32; Mark 15:21; Luke 23:26). The lone dissenting voice is found in John's gospel where Jesus carried the cross himself (19:17). And during the crucifixion, Matthew and Mark agree that Jesus was offered something to drink—a mixture including wine—but he declined to receive it. Luke is ambiguous on the issue. Here Jesus is offered wine vinegar during the ordeal, but the author makes no comment as to whether Jesus accepted it or not (23:36). John's gospel differs once again, for there Jesus both states that he could use a drink and elects to quench his thirst (19:28–30).

The gospel of Luke also differs from the book of Acts (both allegedly written by the same author) when it comes to the occasion of Jesus' ascension into heaven. Whereas in Luke 24:51 the ascension takes place within a day of the resurrection, the account in Acts 1:3, 9 gives Jesus a forty-day period between resurrection and ascension. Or were there multiple ascensions with a visitation in between?

Another example which fails to live up to the standards of corroboration is the two accounts of Judas' death. In the depiction given in Matthew 27:3–5, the penny drops for Judas, who, in recognition of the fate of Jesus is "seized with remorse" at the course of events he had set in motion. Perhaps Judas' intention was simply to get the ball rolling in terms of expediting the arrival of this kingdom of God that Jesus often talked about. Maybe Judas simply wanted to force Jesus' hand into performing a miracle or taking on the mythical mantle of king David and become a conquering hero by reclaiming Jerusalem from Roman occupation. This, after all, was the messianic expectation at the time, that the anointed one would restore the land into Jewish hands, and Judas was growing impatient. Yet things did not turn out this way and Judas' perspective on Jesus' mission was misinterpreted.

Judas might have reacted with astonishment that the imagined messianic script was not being followed. Jesus was not supposed to suffer condemnation. So he returns to the priests the blood money that he received in exchange for his betrayal and commits suicide by hanging. That is one portrayal. The other is found in Acts 1:18–19, where nothing is mentioned about Judas' change of heart. There he experiences no self-recrimination but uses the proceeds to purchase a field (for what purpose we are not informed). Nevertheless, it seems like his wickedness recoiled on him, for while in this field "he fell headlong, his body burst open and all his intestines spilled out"(NIV). A very different story.

Nor are the two compatible. One cannot both return the reward, later to be used by the elders and chief priests themselves to purchase a field, and then also use those same resources yourself for a similar purpose. No economic theory would allow it. If I learned anything from my parents, it is that your money can only be spent once. Furthermore, the type of death Judas is said to have encountered also militates against known physical laws. One cannot both hang oneself and fall headlong with one's internal organs then jumping ship. Natural laws will not permit it. For all its faults, gravity is nothing if not dependable and consistent; its action can be counted on and sometimes annoyingly and inconveniently taken for granted. Correspondingly, one could not fall headlong after being hanged, for gravity would yield a vertical drop with feet landing first (assuming a short distance). Nor I suspect could one marshal the strength to hang oneself once one has become eviscerated. Hence the two tales are mutually exclusive and cannot legitimately be reconciled or harmonized.

One final example in our selection of discrepancies and a minor one at that. The New Testament book of Hebrews claims that the way of all human flesh is to die once (9:27), but in so doing the author overlooks three counterinstances. Neither Enoch (Genesis 5:24) nor Elijah (2 Kings 2:11–12) experienced death, and Lazarus (John 11) suffered the ignominy of enduring it twice, the second time perhaps violently (John 12:9–11). And on the latter topic, parenthetically, the chief priests became envious of the presence of Lazarus, since it was because of Jesus having raised him from the dead that some Jews wanted to see the risen Lazarus for themselves and would then join the Jesus camp. For this reason, the priests wanted to put a stop to the loss of parishioners and devised a plan to kill Lazarus. It seems to be a human reaction to want to be rid of obstacles perceived as in the path of the fulfilment of one's aims. Should those obstacles be people, the desire to accomplish the task by conspiring to kill them has a long pedigree, from ancient times (Julius Caesar) to more recent ones (multiple Kennedys). This does not speak well of human nature.

On a further side note, if the episode about Lazarus was of such grave importance (no pun intended), then why is it found only in John's gospel? I cannot fathom how, if the event actually occurred, it failed to make the editorial cut for the other three gospels.

Back to Jesus and the gospels

Returning to the themes of the gospel interlude, we expand on the topic of human elements in the reporting and selecting of gospel material.

We are slaves to our commitments, and our presuppositions affect what we see and how we interpret it. Even the subject of the date of authorship of biblical books can arouse heated discussion, for philosophical positions about which there is emotional investment are at stake. If one is convinced, for instance, that prophecy can mean more than forthtelling—an exhortation to heed God's call, and can also mean foretelling—a dire warning of usually disastrous events about to occur should the above call not be heeded, then this will influence the stand taken on the dating of documents. Those who accept foretelling will likely deem a prophetic element in the gospels as indicative of a time of writing prior to the foretold events, else it would not be foretelling. Veracity is in the eyes of the beholder. No amount of counterargument could dissuade those committed to interpreting prophecy on the issue of forthtelling versus foretelling. The two camps are at loggerheads.

As concerns the gospels, Mark's foreshadowing of a destroyed temple in chapter 13, a section known as the little apocalypse, must be authentic foretelling for it to have the impact that it does, according to those committed to this view, hence a date of writing prior to the event is non-negotiable. And if its prophetic passages are actually forthtelling, then it would originate after the temple was destroyed. Common ground here is scarce. There is a difference in kind between the two opposing sides, not one of degree, since no matter how much forthtelling there is, it will not add up to foretelling, although this might not be the case for the reverse. This is one way in which positions become entrenched and take on the complexion of zealously guarded territory.

This volume is about human nature and the above describes the way we work at times.

Occasionally the sports fan lives up to his or her name. The term "fan" is short for "fanatic" and it periodically shows. The die-hard fan can even have a den or other living space devoted to memorabilia of his or her favorite team or athlete, a shrine of sorts. In such a temple, no ill is to be spoken of the object of the fan's veneration. If you feel you must, for not every fan has the same affection for the same sports figure, you will need to confine it to your own sanctuary, for there is no room for rivals or detractors here.

Devotion of this sort is not restricted to the contemporary scene. There have been many historical personages who have been held in high regard, then and now. Political leaders can gain a following and many an institution the world over is named after its famous founder. This mentality also extends to religious figures, and icons are not restricted to the Catholic world. Statuettes of the Buddha stand in many a home. And this is not new; they have been around for a while. Buddhists are fans of the Buddha. Manufacturers of these statuettes count on it.

These attitudes were evident in the ancient world as well. A leader who makes a name for him- or herself by vanquishing an enemy or otherwise enabling a nation's citizenry to take pride in its history is well on the way to becoming larger than life. "No one, no mere human could do what this person does," so it was thought. "We must be in the presence of a superhuman. Maybe this person is descended from the gods and hence divine." I speak, of course, of the view of the Egyptians toward the Pharaohs (or did you think I was referring to someone else?).

The Bible can also be treated as a product of these types of sentiments. When the Israelites were struggling to establish a nation, they looked for a king to make them conquerors instead of the conquered. Saul, the first king of the Israelites, did not fit the bill. But then along came David. His campaigns met with success, so it was alleged that God was with him, blessing his exploits. He was hailed a champion, a protector of the realm and defender of the faith, anticipating the British monarchy. David was greeted by cheering crowds and many adoring fans of the type mentioned above.

How better to market another figure in Jewish history? For those who wanted Jesus to appear as a prophet or priest, the best strategy would be to advertise him as a new Moses or Aaron or the suffering servant of whom the prophet Isaiah spoke. Or for those who intended to paint Jesus as a conquering king, introduce him as the lion of the tribe of Judah in the Davidic tradition. Is it just a coincidence that Joseph was age thirty when he entered the Pharaoh's service (Genesis 41:46), David was thirty when he became king (2 Samuel 5:4), and Jesus was thirty when he began his ministry (Luke 3:23)?

To place a figure in the best possible light, highlight only the most exemplary features of this person's life. If one concentrates only on the positive, should there be anything else to report, then the account might circulate that s/he can do no wrong. Before long, the notion that this person is best portrayed as without stain or blemish could be entertained, whether warranted or not. This is how doctrines gather momentum. In order to accomplish this, a strategy must be put in place. For Jesus to be untainted by the sin that beset Adam and all his descendants (as an Augustinian would contend), he must be born without it. To achieve that, in turn, Jesus' earthly journey must be free from that to which we all fall victim—a birth within fallen humanity and God's curse on creation. So we will need to make Jesus' mother likewise outside the fall and curse, and his conception a divine and not a human event. Yet upon reflection this creates additional problems, since how did Mary come by way of the unblemished status? From both parents? (It was not recognized at the time that women have a biological contribution to the makeup of offspring.) If so, then by extension she must have had an unbroken lineage of stainless forebears traceable back to Adam, in which case there would be no need for Jesus' saving work. At the very least, Mary must be an alien to ensure that her seed is as well.

Tricky isn't it? This is some of the difficulty that theology can result in. The trouble is that the gospel writers have an agenda and it comes through in their accounts. No reporting is without it. It might be a noble agenda, but it is no less an agenda for it.

Allow me to come at this from a similar angle and examine what some Christian theologians insist that Jesus must have been like. A typical doctrine about Jesus in conservative circles is that he lived a sinless life. Does this view arise out of reasoned reflection or is it simply a doctrinal requirement? In the Old Testament tradition, what a sacrificial animal had to fulfil in order to assume the role of a means to atonement, or reconciliation with God, was that it be free from physical defect (Numbers 28:31). It could then act as an atoning sacrifice through the shedding of its blood on the altar (Leviticus 17:11). Animal substitutes, though, could not "close the deal"; they could not fully atone for human misdeeds since they did not represent humanity but merely stood in for them. As a result, the exercise needed to be repeated on an annual basis.

The Christian world wanted to adopt this idea of atonement and added a moral component to it. For Jesus to fit the bill, he needed to be fully representative, that is, be like us, and be free not just from physical but from ethical defect. (There is little indication of what Jesus actually looked like, but if he was as festive as the accounts make out, namely, he consumed food and drink, as the Pharisees claimed, to excess (Matthew 11:18–19; Luke 7:33–34), then he could have been a little pudgy.) Moses could not assume the role since he committed murder (Exodus 2:11–12); nor could David because of his adultery with Bathsheba and conspiracy to kill her husband, or at least placing him in a fatal situation (2 Samuel 11 & 12). Even Abraham lied about his half-sister also being his wife (Genesis 20:12–13).

Jesus, in a topic to be expanded upon below, might also have had the "four f's": character flaws, frailties and fragilities, perhaps even foibles, but for the doctrine to hold he must not have had a "fifth f," namely any failures. The trouble with this doctrine is that no ordinary human could match up to the requirement of the law so as to provide an atoning sacrifice without defect. If Jesus was some type of superhuman, then he could fulfil the law, but he could not be our representative. None of us is like this, so how would he be like us? Jesus might be the supreme human exemplar, yet is it possible to be truly human without also being above it? On the one hand, in order to be our substitute Jesus would need to be superhuman; on the other, if he is like us and hence our representative, then he is not without defect and thus cannot satisfy God's atoning stipulations.

The doctrine has good intention—its heart is in the right place but its head is not, for Jesus either takes on too little or too much. If he is without blemish, then that which enables it to be so catapults him outside the ranks of humanity. If he were to have been born superhuman or divine, then he did not get it from either parent and likewise fails to be like us, for we lack any such experience. We know nothing of a stain-free pedigree. And evidently simply being human is not enough. So poor theological Jesus is in a bind: he must be human but dare not merely be human, and must be superhuman but dare not merely be superhuman. And if he were to be both, then he is radically dissimilar to us.

My conviction is that it is sufficient for Jesus to have been God's anointed, the chosen one, in order to fulfil the role of Messiah, as one who inaugurates the rule of God's kingdom on earth. Task accepted and mission accomplished. The job description of Messiah need not call for spotlessness as much as devotion. More on this presently.

Rumor has it that news travels quickly, particularly when it is of the negative variety. So beware of the bad review. Theatrical presentations are adversely affected by them, as are the reputations of public figures. Former President Bill Clinton will forever be remembered for the scandal of his indiscretions, potentially cancelling out, in the eyes of some, the beneficial aspects of his administration and his humanitarian work subsequent to it. As we alluded to earlier, like the black stain on a white sheet, attention comes to be focused on it no matter how large or white the remainder of the fabric. Regardless of the amount of good that is accomplished, a little foolishness tends to outweigh it (Ecclesiastes 10:1).

The reputation of some people, it is claimed, precedes them. This means, among other things, that an opinion has been formed for certain personalities even before they make an appearance. And where people have a hand in someone's defamation of character, the accusation is difficult to erase. Even if someone were to be exonerated of a criminal charge, the proverbial well would still effectively have been poisoned. The default response on the part of people would be to hold the accused in suspicion for at least having been charged. Even if cleared, they might continually be cast in a negative light. This may tell us more about our own human nature than the one on trial.

Especially during political campaigns, it is the task of some characters in one camp to uncover the dirt about the candidate(s) in the other. The more that can be unearthed, the worse it becomes for one's opponent, and the greater the likelihood that votes will swing to the less sullied competitor. Finding these unsavory details is usually not too difficult. We all have skeletons in our closet and some cannot easily remain hidden. Even Mother Teresa had her critics.

It strikes me as interesting that there is at least one historical figure about whom little bad press arises. This Jesus of Nazareth fellow definitely had his gainsayers, for most of the religious authorities of his day vehemently disagreed with his message, largely because he did not meet their expectations. They anticipated a triumphant king in the Davidic tradition to bring them out from under Roman rule, not someone declaring that God's kingdom is established through a returning to God from personal and corporate exile, a captivity to false gods. (In addition to being vilified for what he said and did, he was also maligned for what happened to him, not only for suffering a shameful, even cursed, death, but for [the accusation of] having a scandalous birth.)

Nor did the Romans take much notice of him, that is until the Jews made them notice. The religious leaders detected what they considered blasphemous in his message, namely calling God his "Daddy," yet they could not do away with him through their own legal means, for they had no jurisdiction to institute capital punishment while the Romans were on watch. So they sought a charge of sedition against him. This is a serious accusation, one which the Romans could not ignore, and it was reluctantly upheld. It proved to be an effective strategy (Luke 22:66–23:25).

The religious authorities took it upon themselves to uncover the dirt on Jesus but could not make any charges stick (Matthew 26:59–60; Mark 14:55–56). They had every reason to find the goods (or in this case 'bads') on Jesus since they did not want the citizenry to side with him. They wanted to squelch his ever-rising popularity. The more disciples for Jesus, the fewer 'parishioners' for them. The polls were showing gains for Jesus, so now was the time to put a stop to it. Their own spin doctors would assist in settling the dust afterwards (Matthew 28:11–15).

The best negative press that the chief priests and their cronies could muster was that Jesus had a demon, he was out of his mind, and he ate and drank too much (Matthew 11:18–19; Luke 7:33–34; John 8:49, 10:20, for example). Also that his message was subversive. (Admittedly, being in league with the devil is a serious charge.) Nor did they like the company he kept, for to have common cause with those who were lax in following the Law was perceived as potentially sharing in or becoming contaminated by their impurity and defilement. The worst that others around at the time might say about him is that he created a stir and had the odd temper tantrum (the clearing of the Temple episode), but this is hardly character assassination. There still appears to be little external evidence that amounts to more than hard-core innuendo.

As a side note, the scriptures, in their defense, are surprisingly candid and do not always resort to hero worship, for even the champions of the faith come complete with reported drawbacks. Among them, in addition to ones already mentioned above, are Reuben's sexual indiscretion, for he defiled his father's bed (Genesis 35:22; 49:4); Simeon's and Levi's vengeful acts, for they cruelly retaliated by killing many men (Genesis 34; 49:6–7); Solomon's religious accommodation and compromise to his foreign wives (1 Kings 11:1–10), and so forth. Outside of Jesus, no biblical character could be accused of unflinching, unflagging, undiluted devotion to God, something his four f's could not undermine.

So did Jesus live such an exemplary life that little or nothing untoward could be levelled against him? There have been others touted as exemplars, such as Gandhi, but even they had their hiccups. Can fault be found with Jesus? Perhaps he deserves a hearing while we are preparing our legal briefs.

Honest to God

Another misstep that humans commit is the penchant for passing themselves off as those they are not. We lie about ourselves. When this occurs, it is often for purposes of impressing someone who would be less than impressed if they knew the real us, or for reasons of evading responsibility and the blame and shame that come with it. Bart Ehrman (2011) reveals that a similar infraction obtains for some biblical texts, and if blame is earned for it elsewhere, then this sacred text should not be immune to the charge either.

Ehrman notes that, in the past, biblical books have been excused for having names of authors that likely had little if any part to play in their production. If this were to have occurred in a non-biblical context, there would be no hesitation in indicting the offender with forgery. Ehrman's point is that if someone other than the stated author wrote a biblical work, then s/he (or they) is an imposter and the work should be deemed fraudulent. Yet some have attempted to soften the accusation because, apparently, these authors were honoring the stated authors by affixing their name to it or were writing in the tradition of their masters, to mention two of the most common evasive maneuvers. But Ehrman insists that the charge of forgery cannot be dismissed so easily, neither then nor now.

That the apostle Paul, for instance, was not the author of all the essay-length letters that have been attributed to him has been suspected for about two centuries. His name appears on thirteen missives, while only seven have convinced biblical scholars as being authentically Pauline. This implies that the other six are fraudulent. Furthermore, this raises a difficulty concerning the canon, namely what is considered to be an authoritative body of texts. Having uncovered forgery, we must now ask, authoritative on what grounds? Certainly not always on the basis of its authentic origin.

Picture this. When the texts of the New Testament were being compiled and given canonical status by the end of the fourth century CE, those (men) in charge took on good faith that the stated authors were the actual ones. There are other books, such as the four gospels, which have no stated author, so a name was assigned to each at a later date, both for ease of reference as well as an air of authority. If the name can be traced back to a disciple or eyewitness of Jesus, so much the better for credibility's sake. For the church and its growth-oriented agenda, it's good for business to demonstrate such a pedigree. But this activity is not forgery on the part of the author, for s/he never actually claimed to be Matthew, Mark, Luke or John. The attribution is simply a false or mistaken one. Admittedly, the original faith communities had the agenda to promote their own perspective on the fledgling movement, so they received and used those gospels, out of the many in circulation, that best represented their "take" on the matter. They may even have commissioned the writing of it. This results in a feedback loop where their views were intended to be set to writing and then applied to their communities, which in turn lent credibility to their views. The former corroborated the latter and *vice versa* in an upward spiral. This strategy might not be entirely forthright, but it cannot readily be termed fraudulent.

The dishonesty occurs when books are passed off as having been written by someone other than the stated author. This makes the task of theology difficult. One issue is whether these works should ever have been accorded canonical status. That they have been dishonest about their authorship is one mark in their disfavor. Moreover, as a group they might be giving mixed messages. The assumed real Paul of the seven Pauline letters sometimes writes conflicting statements with those of the believed fake Paul of the other six. Can we have confidence in both sets of statements? Should they be given equal weight? If not, then do we dismiss the fakes and relieve them of their canonical duty? Yet they are all in the canon now, which has been ruled as closed, that is, never to be tampered with. And, it must be confessed, they have contributed to the nurturing of the flock by, so it has been perceived, providing spiritual nourishment. But is this the main or the only concern? If God's Spirit, whose function is to lead all into truth (John 16:13), is pleased to employ all these writings for the spiritual welfare of the church, then the Spirit is okay with mixed messages.

Here is a case in point. It is also alleged that the author of the two short letters attributed to Peter was not actually Peter. Combine this with the fact that when the New Testament uses the term scripture, it has in mind the Hebrew writings, what Christians call the Old Testament. The trouble in 2 Peter 3:16 is that it uses the term scripture in a self-referential way. Here the author writes about the letters of Paul and how some individuals distort them, "as they do the other scriptures." In saying this, the author of 2 Peter claims equal canonical status for both the Jewish Torah and Pauline letters. Was this hasty? Was this how the originally intended audience of the Pauline letters would have perceived it? Was this how the actual apostle Paul, who died prior to the composition of the work, would have seen it? Difficult to say. But now that 2 Peter has been welcomed into the canonical fold, it's too late. Paul's letters must be ranked as scriptural, and so too must 2 Peter by default. Did calling Paul's writings scripture influence their inclusions into the canon? Hence the author, should this have been his or her intent, of a forged document attempts to legitimize other writings of its type and perhaps even itself by implication. Can we accept this as a validation? If so, then good strategy. Kudos, I suppose, for its effectiveness.

Besides, if we cannot trust authors to be forthright in terms of identifying themselves, then can we automatically rely on them for any matter? If the author's identities are forged, then are not the contents equally suspect? Could a similar thing not also have occurred with some statements attributed to God or Jesus? In essence, some authors might claim that "these words would honor God or Jesus, so let's include them." Yet is there honor in forgery?

We further tend to be selective and accord greater weight to those parts of scripture that we deem more important to us. Those who uphold the Jewish Law, the Torah, might also wear fabric woven from blends of cloth, in direct infringement of part of this same Law. Leviticus 19:19 and Deuteronomy 22:11, parts of the Law, state (specifically the latter) that we "shall not wear clothes made of wool and linen woven together." To do so is a serious wardrobe violation which we are content to ignore. We overlook the inconvenient, perhaps when the rationale of such laws escapes us (Jacobs, 23-25). By conferring greater divine approbation on our pet passages, we excuse ourselves from infractions committed through what we wish to regard as lesser injunctions, and we do so by that same authority.

Moral of the story: the writings of this sacred text, those represented in the canon, are based on human decision on a community scale. Critiquing the Bible, though, need not reduce to a salvage operation geared toward distilling or sifting the contents for an undistorted residue. Rather, it can be a program of discerning what wisdom can be gleaned from the accounts for our current situation. This appears to be a more salutary approach than centering on the fragments that remain after the contemporary materialistic critical mindset has sanitized it.

PART TWO

Historical Material

HAVING SURVEYED SOME BIBLICAL RESOURCES for information pertaining to human nature, we now turn to a more theoretical look at the topic. To this end, a very useful text for our purposes is *Ten theories of human nature* by Leslie Stevenson and David L. Haberman (fifth edition). Herein the two co-authors offer a short summary of four ancient religious traditions, two ancient and three modern philosophical systems, plus a concise outline of several others and lastly rounding out the discussion with Darwinian theories. The authors adopt the same formula for each treatment: an examination of any metaphysical themes that need to be addressed as background for each system of thought; an investigation of the perspective on the human condition under each tradition; a description of the predicament that humans find themselves in—giving us a diagnosis as to what has gone wrong; and finally a prescription of what might constitute a remedy for repairing the problem together with a prognosis of its success if endorsed or danger if left unattended. The authors do not ostensibly submit any approach as a clear favorite, but tender historically significant options and allow the reader to decide what might be the best course of action from among the alternatives. They also encourage open-mindedness throughout the process.

Confucius

Stevenson and Haberman begin their analysis with ancient religious traditions and devote their opening selection to the thought of Confucius (551–479 BCE). K'ung Fu-tzu, or Master

K'ung as he is known, does not dabble much in metaphysical matters—that which is beyond the ordinary human sphere of access—but concentrates on human welfare and proper conduct. To this end at least, the beyond does play important roles: Destiny, a force which governs our station in life as well as our life expectancy, lies beyond human control; and the Decree of Heaven, a force which governs morality, lies within it. At best we can only accept the former as our fate, but we are called upon to abide by the latter. Determinism dictates the first, free choice the second. Heaven urges that we engage in "cultivating a transcendent morality," yet we possess the power either to conform to or reject the decree. This mandate is directed to everyone and Heaven supports those who seek to be in harmony with its decree and work toward ultimate perfection.

The trouble, however, is that we tend not to conform. Those who do are known as sages and we are to follow their example. The likelihood, though, of achieving success in this pursuit, given human nature, is meager. Confucius himself does not expect ever to come across one who adheres to this Way. This is due to five ills that confound and trip up our efforts: we are greedy and selfish because we are self-absorbed and -interested and attached to personal gain; we show disrespect to family members because we have lost filial piety; we tend to be hypocritical; we are ignorant of the Way of the sages; and we lack benevolence. The most important virtue to nurture is benevolence and its primary relationship is that between father and son.

Confucius is optimistic about the human potential to follow the way of the sages, but at the same time is pessimistic about our rising to the challenge. Apparently we have it in us to attain ultimate perfection, but the prospects are grim. The way forward for Confucius is to study the Classics, the repository of wisdom from the past. The path of moral perfection involves the practice of the rites, that is, "ritually correct behavior" as taught through the Classics, which in turn reveal the way of Heaven with the sages as its exemplars. In essence, these observances yield the Golden Rule, sometimes stated in negative form as: "Do not do unto others what you would not have them do unto you" (Stevenson & Haberman, 10–21), whereas the gospel writers portray Jesus as framing this in the positive. Put bluntly, Confucius does not anticipate ever coming across a sage. Perhaps this makes Confucius a realist when it comes to human nature: the possibilities are there for perfection, but few will avail themselves of the opportunity.

By way of commentary, Confucianism is very much a this-worldly undertaking, making, as it does, no reference to God, the soul or an afterlife. The conduct and perfected inner state which Confucius advocates is purely for human welfare in this life, that society might truly be social and civilization civil. One must ask, though, if this is adequate for those holding to personal existence beyond death for which this life might be considered preparatory. Would one live a life in precisely the same manner as Confucius' recommendation if an afterlife were not in view? The motivation would likely be different if derived simply from concern for the welfare of humans in this life. Moreover, could this tradition be considered a religiously satisfying one; does it lead to a fulfilling life and meet the spiritual needs of its adherents? As with the previous question, it supposedly depends on the expectations one brings to it. Further, is religious observance to be understood as equivalent to a code of conduct, as the above suggests, or something more? And lastly, does the beyond ever break in to the here and now? Questions like these will occupy our interest in the pages ahead.

Hinduism

Stevenson and Haberman's second look at human nature through the lens of a religious tradition comes from Hinduism. Since there are a variety of schools of thought within Hinduism, the authors opt for the Upanishadic version as an adequate cross-section of the tradition. Hinduism, virtually alone among the world's major religions, is without a specific founder and bears multiple sacred texts. Together with the *Rig Vedas* and the *Bhagavad Gita*, the *Upanishads* (7th–8th century BCE) form the main corpus of Hindu scriptures.

Among the concerns addressed in the Upanishads are metaphysical themes, with particular attention directed toward the nature of reality. A major issue therein is the perennial question of the one and the many, that is, what is the one that unifies and interconnects the multiple forms that we find in our ordinary everyday world of experience? Upanishadic Hinduism is an early attempt, even prior to the Greeks, at a resolution to this problem.

The Hindu worldview can be outlined in the following ten-point summary:

1. The ultimate being is the One or the All, and this divinity is named Brahman. The debate continues as to whether this God is personal or impersonal. Whatever the case, it is both immanent and transcendent, in the world and beyond it.
2. God divides God's self into a multitude of selves (including us) and the world of myriad forms. Since the one self produced many selves, this becomes the answer to the question of the one and the many, namely Brahman is the principle which unifies the many.
3. From the above, our self originated from the ultimate Self and our inner self is referred to as atman. Our true self is immortal and eternal since it came from God.
4. Extending the previous point, our true self is part of God's true Self, hence atman is actually Brahman—a minuscule part of the One. Atman is the Brahman in each person.
5. Our predicament as humans is an identity problem, specifically we are ignorant of our true nature. We assume that we are the extent of our external ego self.
6. The remedy lies in coming to know our true inner self. Thus salvation is based on what we know or come to know.
7. This can be accomplished through yoga and meditation.
8. Failure to do so fixes one onto the cycle of birth, death and rebirth, though the Upanishads are unclear as to whether reincarnation is a positive or negative occurrence.
9. A necessary co-requisite to seeking and attaining liberation, that is, breaking free of this cycle, is extricating oneself from attachment to the world of objects. Karma is the universal principle built into the system that apportions to us our just desserts, meaning we get what we deserve. Karma either keeps us on this cycle or allows us to be released from it. The choice is ours.

10. If we understand the world as made up entirely of transient objects, then we will suffer their same fate. Since we are immortal and eternal, however, death is not the end and we will return to make another attempt at liberation (Stevenson & Haberman, 27–36).

While our volume is not an exercise in comparative religion, I do wish to mention two ways in which Hindu ideas differ from those found in conventional forms of Christianity. First, whereas in Christianity the world is composed of matter, which is a substance different from that of the divine makeup, in Hindu thought the world is formed from God's Self, God's own being. Second, in Hinduism to truly know yourself, which is to know your true inner self, is to know God's Self. On the contrary, in Christianity to know yourself does not automatically bring you any closer to knowing God. The reason for the second point is found in the first: there are at least two substances in Christianity, only one in Hinduism. A standard Christian theological distinction is that to know matter is not necessarily to know spirit. Having said this, admittedly Christianity is also based on what one knows. Like in Hinduism, though for a different reason, knowledge of God is salvific.

Buddhism

The authors next focus their attention on a consideration of the Buddhist worldview. The two main traditions found here are the Theravada or Hinayana school and the Mahayana version. The former is known as the Lesser Vehicle, in that only certain persons may apply, while the latter is referred to as the Greater Vehicle, for it is open to all people. The former is represented in the scriptures of the Pali Canon, whereas the sacred text of the latter is the *Lotus Sutra*. Hinayana Buddhism pictures a human Buddha, named Siddhartha Gautama (563–483 BCE), who encounters struggles, becomes enlightened and passes his teaching on to others. Mahayana Buddhism, in contradistinction, reveals a transcendent Buddha who is cosmic and eternal, who works salvation, accepts worship and is called a savior. The latter's visitation to earth is in avatar or "phantom" form.

Buddhism grew out of Hinduism and retains some of its forerunner's terminology. Buddhism accepts the idea of karma, reincarnation and attachment. The concept of karma undergoes little alteration, reincarnation (or more precisely, rebirth) occurs not only from one life to the next but within the same life in a rapid-fire way, and attachment becomes the idea of craving and grasping onto what does not last and from which we need to become detached. On this very topic, the Buddha embarks, in his inaugural sermon, on a diagnosis of the human condition. He assesses the basic human problem as one of, as the Rolling Stones would concur (though with a different emphasis), dissatisfaction. We ache with the knowledge of the impermanence of every material object that we value, ourselves included. We would rather that everything we hold near and dear be permanent. And the things we do regard highly are not without their faults, that is, they are never quite right and this produces anxiety and insecurity. We also fit into this category. Further, we only imagine that there is a constant self from one moment to the next, but the Buddha says no. He claims that our impression is deluded since we possess no "I" or soul or autonomous self.

This marks the first of four noble truths, that permanence is nowhere to be found, the recognition of which causes us grief. The second is that our predicament boils down to wanting to hold on to the constantly changing and make it permanent. Rather than accept the transitoriness and interconnectedness of all things, we try to forestall transience. Much to our dismay, the attempt is futile. So what do we do about it? The Buddha informs us of the third noble truth which assures us that the first two can be overcome. The drive for grasping can be extinguished. Specifically, "When craving becomes extinct, then [dissatisfaction] is gone, and the result is [the blessed state of] Nirvana." This healthy state is achieved through the fourth noble truth, which indicates that there is a better way, namely the middle way or eightfold path that leads to Nirvana. These eight tasks involve proper approaches to ethical conduct, mental discipline and wisdom. The eighth in the series, coming full circle, is the right understanding of the four noble truths (Stevenson & Haberman, 47–64, 68).

A well-worn objection to the Buddhist proposal is the alleged contradiction of the problem of craving and its relation to Nirvana. The difficulty is presented this way: Nirvana is a worthy goal to pursue since its result is the end of all craving. Put in other terms, it is a pursuit to end all pursuits. Yet how, the critics maintain, can one strive for that which, or the goal of which, is the end of all striving? One is supposed to arrive there by surrendering the penchant for striving, not by enlisting it; grasping is to be relinquished not conscripted. One way, I imagine, of extricating oneself from the circularity is to counter with the point that the dissatisfaction one encounters in not yet having attained Nirvana becomes a spur, an incentive, a motivating factor in continuing to press toward it. Not all goals are unworthy of the pursuit. Fixing one's gaze upon a goal is honorable if the goal is commendable. With this type of reasoning, Nirvana is an admirable aim.

One difference between Buddhism and conventional Christianity is that whereas the former emphasizes non-desire, the latter stresses right desire. The proper desire is to follow God's Messiah and become more like him. Yet as was just mentioned, it could be argued that the Christian's desire is parallel to the goal of the Buddhist to attain enlightenment. Both could be said to be passions, though both Jesus and the Buddha were passionate about being non-passionate, or not being led astray, regarding what leads to gratifying selfish desires. As the apostle Paul claims, a point to which we shall return below, there should be a turning on the part of each individual from the old nature to the new. This appears to be a legitimate desire that one could be passionate about. Perhaps both Jesus and the Buddha would approve.

In like fashion, while Buddhism focuses on the release of suffering, Christianity proposes that a certain type of suffering might be called for. The call on the part of the Jesus of the gospels is for us to accept the suffering that comes with following him, known as taking up the cross and participating in his sufferings. Discipleship in Christ is costly, for finding our life means losing it for Christ's sake. The Buddha would applaud the sacrifices one makes for the sake of enlightenment, for it could be said that this comes with suffering. To relinquish desire is to suffer loss and to be tied to impermanence is also to experience loss. To turn from impermanence is to lose a transitory comfort. This constitutes a sacrifice, though both Jesus and the Buddha would agree that it is a necessary part of the process. No psychological pain, no religious gain, I suppose.

A Christian might be inclined to object that the Buddhist diagnosis about the main human dis-ease of suffering is wide of the mark. The basic human problem indeed begins with us, but it is not so much about what happens to us as what we perpetrate. Buddhists see our perception of the world in unsatisfactory terms. The problem rests with us, for we sense that somehow we have gotten short-changed. The world is not quite up to what we desire it to be. Christians, on the contrary, understand that our impressions of the world and all its faults are secondary. The trouble is we have made bad moral choices which might recoil on us. The problem rests with others in that we may have infringed upon their rights or dignity and they are giving us a poor review. For Buddhists, the issue is more about us and our perceptions; for Christians, the concern is more about others, our neighbors, and our relations with them. Both Christians and Buddhists would concur that each other's diagnosis is partially correct but that it has not accurately targeted the fundamental predicament. Together, perhaps, all the bases are covered.

Moreover, thinking from a wider perspective, our pursuit of release from suffering might come at the cost of increased suffering on the part of others. Even inner attitudes can have external consequences, for everything we do has wider implications. If it takes us away from relational obligations, seeking personal, individual liberation can produce adverse effects. The Buddha himself abandoned familial commitments—a wife and a newborn (p. 51)—in favor of his quest for enlightenment. If interviewed, would they give Hubby/Daddy a positive review? If there is emotional blood on the floor, then maybe the price is too high. (One wonders if they could not have become willing participants, even partners, in Dad's own quest.) To be fair, however, it must be admitted that the Jesus of the text even counseled others that to be true followers of his might entail leaving family for his sake, the members of which are to be "hate[d]" by comparison (and the same goes for the devotee him- or herself) (Luke 14:26).

One additional item before we turn our attention elsewhere. On the topic of impermanence, do the Buddhists have it correctly that everything is in flux, that nothing in the world ever remains the same? In physics there are what are known as the physical constants of nature. These include values for the speed of light in a vacuum, the rest mass of elementary particles, fundamental electric charges, and the gravitational constant. And they are not supposed to change; they are constants after all. Yet some astrophysicists inform us that they are not entirely constant. Some values might have changed, are changing or will change. The modifications may be minimal, sometimes a mere seven parts in a million, the amount of difference that earlier equipment could not detect. But the issue is not the degree of change but its actuality. If true, then constants do not live up to their billing (Barrow (2002)). Constants would need to be constant in order to constitute permanence, and if they cannot, then the Buddhists may be on to something.

Perhaps it is not possible to point to something in this world that undergoes no change. But might permanence be the property of another realm? The ancient Greek philosopher Plato takes up the challenge and urges that eternality and temporality can be combined.

Plato

With Plato (427–347 BCE) we move from the world of Eastern religious traditions and enter the arena of Western philosophy. Plato maintains a dualistic distinction about the world on at least two fronts. First, reality is divided into two things: elements and compounds. Elements never dissolve into component parts, hence they are always the same, eternal and unchanging (in every way except for locomotion). Compounds, on the contrary, always proceed towards dissolution, implying that they are never the same. This state of affairs prompts the famous phrase "change implies corruptibility," meaning there is nothing, if it undergoes change, that does not eventually experience corruption. This makes Plato evaluate the static highly and the changing lowly. Second, the human is likewise divided into two, known as dipartite: body and soul. The body is a compound which never ceases to change, and for this reason is to be devalued. The soul, however, is an element which means it is eternal and of great worth.

Plato's innovation is that there must be a place where permanent elements reside, so he posits a realm of Forms in order to accomplish this task. This is a world beyond the bodily senses that only the mind or soul can reach. It is where the pure and perfect exemplars, or molds, exist, of which all the objects and abstract concepts of our world are representations. Yet they are poor reflections or pale imitations. All forms are eternal elements, like virtues, colors, numbers, and so forth, and this includes the soul. This is where the soul finds its true home and this is where it seeks to return. Periodically, the soul descends from the Forms into a body from which it wishes to be released. I say periodically since Plato entertained a doctrine of reincarnation. The soul desires to be released from its "prison-house of the body" and to reascend to the Forms. Needless to say, to the extent that Christianity exhibits Platonic parallels, it has inherited a negative view of the body as that which hinders the spiritual life. Body bad, spirit good.

The truly real is that which endures; a changing reality is a mere approximation of this truly real. In like manner, true knowledge is afforded only by the truly real; a changing knowledge is not true knowledge. Only the soul, as a member of the Forms, can bring us in touch with reality and a knowledge of it. Plato employs a cave analogy to describe how limited our sight and knowledge are while in the body. We are like cave dwellers chained in position to perceive only the back of the cave and are dependent on the light from the entrance to the cave, or to a fire, behind us to allow vision. Yet all that we see are the shadows that we and other objects cast on the back wall. The extent of our knowledge, then, amounts to shadows. As our souls are incarcerated in our bodies, so is our knowledge confined to a cave. And as souls seek to be released from bodies, so knowledge desires to be liberated from tunnel vision.

Whereas a person is dipartite, the soul itself is tripartite. (Perhaps tripartite is not the best term here, since the soul, as an element, cannot have parts. It does not become a compound by residing in a body, for then it would be changing and hence no longer eternal. It would thus ultimately succumb to dissolution and no longer be in a position to be reincarnated. Consequently, it might be more precise to refer to souls as containing aspects or principles as a way of circumventing the difficulty.) One side of the soul is rational and is called Reason, another focuses on desires, drives and appetites and is termed Irrational, and the final one centers on passion and emotion and is referred to as Spirit. Plato uses another analogy to assist us in understanding how the three work together. He asks us to consider a chariot with Irrationality, as the most prominent and insatiable aspect of the soul, in the shape of a dark horse, Spirit as a good white horse (is this where Hollywood obtained its inspiration for having the good guys wear the white hats and the bad guys the black?—another dualism, this time of light and dark) and Reason as the charioteer, mandated with the task of keeping the two steeds in line. If one were to become unruly, it could make the chariot veer off course. Only if Reason is in control can the charioteer properly steer the chariot. Harmony among the three principles leads to a virtuous life individually and a just existence societally. Rulers who are most fit to govern are those who have a good philosophical upbringing and education, having been trained in the proper use of reason. They are known as philosopher-kings.

We are now in a position to answer our initial questions. For Hindus, Brahman is the One who is the source of the many; for Plato, there is one realm of Forms, or one Form per each concept or object, and many objects. For Buddhists, there is only impermanence; for Plato, all the contents of the realm of Forms are permanent, eternal and unchanging (Stevenson & Haberman, 74–85).

There are several ways that Christian thought has drawn from Plato's work. In the final section (chapter 40) of the final book (10) of the final part (6) in Plato's *Republic* lies the Myth of Er—an odd way to end a philosophical treatise—with a myth that describes his doctrine of reincarnation. While most forms of Christianity do not accept such a doctrine, there are some striking parallels. Here are a few. The soul originates in or with the divine; it pre-exists bodily birth and survives bodily death; it falls as a penalty for previous sins committed; it faces a post-death judgment; it earns rewards or punishments as a result of just and unjust deeds; and once purified experiences ultimate deliverance. As Christianity drew from Plato, so Plato may have drawn from Hindu and Buddhist sources about cycles of birth, death and rebirth, as well as karma, in the sense of the effect that conduct in the present life has on the quality of existence in the next. Karma operates also within a lifetime, but Plato does not entertain this.

Further commentary on Plato awaits an examination of his student Aristotle, for the student serves as a partial critic of his master.

Aristotle

Aristotle (384–322 BCE) both agreed with his mentor Plato and made a departure from him. Two ways in which this is the case call for our attention. First, Aristotle retained the use of the term Form, but for a different reason than his teacher. He rejects Plato's idea that Forms exist in another realm while their poor representations inhabit the world. Aristotle disliked this separation. Material objects, he argued, are not instantiations of Forms, but each class of things bears a common property. All of them have it, and those things that are not them do not have it. And having them means that this is where the property resides and not somewhere else. Properties are not separate entities but are found in objects, though it becomes difficult to describe them, such as horseness, without their verging on thingness. Compounding the difficulty is that the application of a property to one thing makes it different from its application to another. The good that is the taste of ice cream, for instance, is different from the good that is a properly working vehicle. Aristotle is left with this philosophical problem.

The second departure from Plato, for our purposes, is Aristotle's view of the soul. Animate things are alive because they are ensouled; inanimate things are not alive because they lack this feature. As with the discussion in the previous paragraph, the soul is not a thing but rather a set of properties that combine to describe who we are and what we do. This is the way we function, our mode of operation, in essence, our resume. This way of living cannot be found in a corpse, for only animated bodies have a way of life, meaning that there is nothing that can exist apart from a body. Only living bodies perform the act of functioning, hence there can be no mind or soul without a body. Thus it is best to describe the soul as a set of abilities or capacities of an animate object. Aristotle runs into trouble here as well, since God, or the gods, appear to function quite nicely without a body. Using Aristotelian logic, first premise: the soul is what makes something animated or alive; second premise: the gods are alive—the gods might transcend ordinary living, but they have at least that; conclusion: therefore, the souls that the gods "possess" or that make them alive can in fact survive without a body.

But there is more to Aristotle than metaphysical speculation. He also comments on the way of living that best befits humans. He accomplishes this by connecting three concepts: happiness, virtue and good. "Happiness," he contends, "is the chief good"; and "human good" is "the best and most complete" virtue (Stevenson, 71–72). If we conduct ourselves in accordance with these ideals, then this will produce an "admirably fulfilled life." In language to which the Buddha would have given his approbation, virtue is to be found between the extremes of vices. For example, "Courage is the right balance between cowardliness and rashness." The extremes are illegitimate since they "fail either by deficiency or by excess." Cowardliness constitutes insufficient courage, rashness an overabundance. This state of affairs is reminiscent of the Buddhist middle way which forges a path that is "a mean between two extremes."

Aristotle urges that the best pursuit on the part of humans is a life devoted to contemplation and intellectual activity, perhaps because this is the very preoccupation on the part of the gods. Yet this leaves us with a perplexing question. Who are those fortunate few who could devote their lives to such an undertaking? Reflection might not be the first thing that comes to mind, say, for a homeless person. The whereabouts of food resources might form their primary focus, maybe even an all-consuming interest. Hence is Aristotle's ideal of an admirably fulfilled life open only to the privileged? Does this make him an elitist? Since not everyone can aspire to this ideal, it will inevitably leave some persons on the outside or on the sidelines. Consequently, Aristotle's message is directed to those with opportunities for such fulfillment (Stevenson & Haberman, 91–105).

Judeo-Christian Scriptures

The Judeo-Christian tradition differs from others in the following way. Whereas the Greek philosophical gods value rationality, in that they engage in intellectual activity, the God of the three Abrahamic faiths (Islam as the most recent) displays not only knowledge but compassion and calls us to do likewise. Our status as humans is all that we require to proceed on the path toward godliness; we need not be intellectual giants or privileged in some fashion. God also empowers us to walk in this way.

There are at least four theological themes that Jews and Christians can agree on. First, humans are created according to something known as God's image or likeness. This does not require our having arrived on the scene in a mythical sixth day of creation, rather we could have put in an appearance through a lengthy evolutionary process. The point is that humans in this doctrine bear something that is unique and qualitatively different from the remainder of the animal world, and that can partially be described as participating in the ethical qualities of the divine. The image of God means more than this, but it amounts to at least that, hence the term ethical monotheism—one God who delivers a moral code and expects us to follow its precepts. More than this, while other animals have their tasks and operate with a drive to secure food, sometimes shelter, but not clothing, humans have been given a mandate. We bear the responsibility, and have been given the duties and obligations, to be caretakers of the world of which we are a part. God has no such custodial or managerial expectations of any other creature.

Second, something went wrong and we disobediently opted for our way of doing things instead of God's. This peremptory self-assertion created a rift between us and God as well as between us and others. Third, God did not scrap God's artistic canvas but creatively sought to restore the broken relationship and reconcile the two parties. Fourth, God attempted to accomplish this through a series of covenants, or contracts, geared to bridge the gulf. The trouble is that these measures could not remove the guilt of our disobedience but only provide temporary relief. Priests must perform the same rituals annually and in perpetuity.

This is where Christianity comes in and the departure from Judaism occurs. What the old means could not achieve, the new supplies. The new covenant, or testament, teaches that the one whom God anointed Messiah successfully erased the guilt. What remains is for us to follow this Messiah, become more like him, and be agents in the program of reclaiming the entire world into God's domain.

Not only is the Abrahamic view of God different from that of the Greeks but so is its perspective on human nature. There is no metaphysical dualism of body and soul here as there is for Plato. In the Judeo-Christian understanding, there is no highly-valued soul which survives and separates itself from a poorly-respected body at the point of death. Instead, bodies are valuable enough to be resurrected and the Messiah gives us the first glimpse into this. The duality that occurs is not between two ways of being but "two ways of living." It is not a matter of good soul, bad body, but valuable person. So the objective is to redirect the mode of living from an old to a new way, from unredeemed to redeemed, from unregenerate to regenerate. This involves a reorientation of the heart or will. The problem is not our physicality but the misuse of our faculties. The trouble with humanity, and this is the bad news, is that we are flawed, we inevitably, though not necessarily, militate against God's intentions for us. Yet the good news is that God considers us worth reclaiming and reorienting. God has witnessed our malady and has provided a remedy. That includes the world and our bodies. Plato would not give his consent to such a project.

Certain sports fans have been known to hold up signs at sporting events with the scripture reference John 3:16 emblazoned upon them. In all likelihood, however, they might fail to understand the implications of the verse. The significance is not just that God used the Messiah to work reconciliation, which is the aspect of the verse which these fans probably have in mind, but what constitutes the object of God's concern. The verse begins with "For God so loved the world," something not to be overlooked, since therein lies the main focus of God's attention. God values the world so highly that God considers the Messiah expendable. The cost of reconciling the world is the gift of the Messiah, a price that God was willing to pay. An expensive gift for a valuable world. The Platonic vision is of a disembodied state in another realm, like the Forms; the Christian vision is of an embodied resurrection in another era or age, on a renewed earth (Stevenson & Haberman, 105–26).

An extended theological discussion about human nature will follow in Part Four.

Descartes and Hume

We now fast-forward from the ancient to the modern world and the first modern philosopher, Rene Descartes (1596–1650), who died on the eve of the scientific revolution. Like Plato, Descartes was a dualist who believed that there were two classes or categories of things in the universe, namely the physical or material, and mental or thinking substances. (Kindly ignore for the moment the inherent contradiction in the term immaterial substance.) The soul, which he understands as interchangeable with the mind, is a separate entity, and in reference to the mind-body problem (to be examined in Part Three), numerically distinct from the brain. That is, there is a brain as part of the body, and another entity known as a mind. The soul can exist apart from the body, but, unlike for Plato, it does not originate from a realm of Forms but from God, who made and deposited it in a body. This means that whereas Plato's soul is eternal, having no beginning or end, Descartes' soul does have a beginning since God created it at some point, though from then on it is everlasting. As a devout Catholic, Descartes believed that God attaches souls to bodies, the former surviving when the latter perishes.

An issue began surfacing at around this time, namely the extent to which the scientific method can be applied to the study of humans. Descartes' position uses the following logic and makes these four assumptions. He reasons first that physical objects can be described mechanically and have mechanical explanations, machines that they are. Next, since they are reducible to mathematical formulation, they are deterministic. This means they can be described using cause and effect and are therefore able to be studied by science as a public undertaking. The same sequence applies to the mind, though with the opposite results. Mind does not have mechanical explanations since it is not a machine; neither is it reducible to mathematical formulas since it can exercise free will; nor can it be described using cause and effect. All of these combine to prompt Descartes to declare that mind is not a subject for the scientific enterprise as mind is private and not open to public scrutiny. (This division, of course, is not so straightforward when it comes to the quantum world, a topic that I tackled in a previous work.)

On the opposite extreme appears philosopher David Hume (1711–1776). Like the Buddha, Hume believed that there is no enduring self to speak of. The sense of self we possess is the impression we obtain when we string our memories together. The truth for Hume is that episodic remembrances do not by themselves forge a continuous self, despite our tendency to believe they do. He applies the same logic to cause and effect: all we ever experience is the succession of one thing followed by another, yet this in no way permits us to claim that the former caused the latter. So too with our sense of self: one memory follows another, but this does not guarantee an unbroken self that experiences them (also a theme to be developed below). As a result, the second sequence we outlined for Descartes is deleted for Hume. If there is no self, then all is material, and by the above reasoning, all comes under the scientific umbrella. Moreover, Hume adopts the working assumption that since everything about humans can be studied by science, this includes what humans do, such as the religious motivation itself. Religion becomes still another item for scientific investigation (Stevenson & Haberman, 137–42).

Freud

One approach in the Humean tradition comes to us by way of Sigmund Freud (1856–1939), a chapter on whom the authors included in their fourth edition but not here in their fifth. (Our inclusion of his thought at this juncture marks the only anachronistic insertion we will entertain.) Given his medical training as well as the intellectual climate at the time, research into hysteria and nervous disorders were believed to point to the idea that since the root problem of physical symptoms could be emotional conditions, then the mental world could influence the physical and even adversely affect it. Materialist and determinist that he was, Freud saw even our mental world as describable in terms of a series of causes and effects. Choices become determined by hidden causes and these must become unlocked. The implication is that we have no free will and are not even in control of our thoughts, words and deeds.

Much like Plato, there are taken to be three aspects to our mental makeup, the largest of which is mostly hidden. All of our dysfunctional features can be traced back to the emotional trauma that we have stored up in this hidden part of us. In fact, it is so well hidden that professional assistance is required to tease it out. A healthy situation obtains when all three aspects work together harmoniously and a dysfunctional one when they do not. We all have a "'defense mechanism' by which people attempt to avoid inner conflict. But it is essentially a pretense, a withdrawal from reality, and is doomed to failure. For what is repressed does not go out of existence, but remains…" (p. 164) and makes its presence felt. Untreated trauma expresses itself in irrational behavior, which Freud termed neuroses. When one stores up trauma in this hidden part of our mentality, we relinquish authority over it and fall ill to mental disorders. Health is restored once we regain control over our submerged painful events.

Historical Material   47

The task of the therapist is to foster the resurfacing of these emotional disturbances. Delving into the source behind such seemingly innocuous actions as neuroses, however, may not come across as harmless to the patient. There arrives a point when free word-association and other techniques might be viewed as infringing on personal privacy once it hits too close to home. Freud suggested that a different approach might be in order. He found that there was less resistance and fewer barriers set up by the patient when dreams are the topic of discussion. The patient feels less inhibited then, for it appears that this is one area where the protective reaction of the patient's hidden mental component lets down its guard. Perhaps the submerged portion feels less threatened because it believes that there is less here to hide. No one ever accused the ordinary mind of being a brilliant military strategist and here it commits a tactical error. Freud then assumes that the divulging of dreams gives insight into the patient's darker side, which, once brought into the light promotes the remedial process. The patient can then face his or her fears, discharge the stored energy and be released from the emotional disturbance. This is the way of salvation for Freud. (Freud also incorporated a modified version of Plato's chariot analogy in his tripartite view of the human mind.)

The trouble which Freud recognized, though, is that we are all in a similar boat, in that we are left with two equally unattractive prospects. Either we conform to cultural standards, resist our drives for instant gratification of desires, submerge them and experience "anxiety and frustration," or we surrender to them and then suffer from guilt. If we follow our duty and do our part in working toward a civilized society, we suffer psychologically. If we follow our urges to the neglect of society, we also suffer psychologically. Either way, mental illness cannot be avoided; society cannot help but produce psychological pain (Stevenson & Haberman (fourth edition), 157–66; Hedges, 117).

Kant

The authors next turn to additional modern thinkers, beginning with Immanuel Kant (1724–1804). In relation to human nature, Kant's thought can be summarized in the thirteen items listed below:

1. Humans act out of self-interest, something with which Hindus, Buddhists and thus not only Christians could all agree. Kant calls this the hypothetical imperative, which involves the gratification of our desires.
2. In conjunction with the above, we also recognize moral obligations, which Kant refers to as pure practical reason.
3. Based on the foregoing, we also act out of duty simply because it is the right thing to do and for no other reason. This absolute, unqualified obligation he names the categorical imperative.
4. Similar to the animals, there are causes for what we do. Part of physical behavior is deterministic and has mechanical explanations. In these instances, humans are like machines.
5. Unlike the animals, there are also reasons for what we do, and in these cases our free will is expressed in the choices that we make, for choices are not caused. As rational beings we act out of reason and hence are free. For Kant, "morality is fundamentally a function of our reason, not just our feelings" (p. 151).
6. Sadly, the human condition is such that we do not naturally do what is right. In theological language, humans are depraved—our intent is often to misuse our freedom to make bad moral choices. We thereby subordinate our duty and obligations toward others to our inclinations.
7. We are naturally evil because we are born with it. It is innate, part of our essential nature and a heritable trait.
8. Evil is in us also because we choose it. Left to ourselves, we perpetrate it. (There is a long tradition of grappling with this issue and includes such luminaries as Augustine. The distinction arose of evil as inevitable (we will sin) or as necessary (we must sin); it is going to happen versus it needs to happen.)
9. Even though we can never attain moral perfection, we do have the capacity to do what we ought. Kant surmises that

"ought implies can," meaning we can marshal the power to do what we should and thereby improve morally.
10. A barometer for ethical decision-making is the universalizability of our actions, that is, could we approve of the same action on the part of all other rational beings in similar circumstances? The most important universalizable maxim is that people should be treated as ends in themselves and never used merely as a means to an end.
11. A system of rewards and punishments might "induce outward conformity to rules," or may fail to be a deterrent, "but it cannot create the virtuous inner attitude, the will to" act rightly and with proper motivation (p. 157).
12. Morality is the basis or essence of religion—its starting point is our sense of moral obligation.
13. Religion is therefore as independent of and separate from science as fact is from value.

Much of the foregoing can be traced back to Kant's Pietistic upbringing, "a radical spiritual movement within Lutheranism that emphasized personal devotion and right living above dogmas, creeds and ritual" (p. 144). Kant would agree with Jean-Jacques Rousseau that "true worship is of the heart" (p. 143). Notice that we have undertaken this summary without any explicit reference to divinity. Even though Kant held that duty comes from divine commands, he further claimed that "faith needs only the *idea of God*" (p. 159). Unorthodox views such as these fuelled the acrimony he received from the authorities in his native Prussian homeland (Stevenson & Haberman, 143–60).

Some comments are in order, particularly on points 7 through 9 and 12. The first concerns the assumption that what we ought to do is also something that we are able to do. The entire history of the Judeo-Christian tradition has shown that humans are flatly incapable of abiding by divine precepts. The God of the text had introduced a series of covenants as a means of providing temporary assistance in covering misdeeds, but they could never eliminate them. Some Christians believe that the Messiah accomplished what previous covenants could not. If God went to these lengths to achieve those ends, then we could not do so under our own steam. This would seem to undermine Kant's assumption. His point is that God would not ask us to carry out a task if we lacked the ability, yet orthodox doctrine states otherwise. Christians of certain descriptions would claim that God wishes to impress upon us the very fact that we cannot fulfil God's law by ourselves. To trust in God and not our own strength is an insight which Martin Luther disclosed and Kant downplayed. For Luther, we can neither do the right nor will it, and we require God on both counts. Kant finds that our power may not be complete, but it is real. In this he departs from his Lutheran tradition.

Shifting our focus to the dilemma about the onset of evil, one resolution is a both-and approach, namely we are both born with evil and we fan it into flame as we proceed. The question then becomes is the presence of evil necessary or inevitable, that is, respectively, must it occur or will it simply take place? Were evil inherent it would be expressed necessarily; should it manifest itself through sheer acts of will, then its appearance arises inevitably. Hence, can something be both necessary and inevitable?

Lastly, we can ask whether or not religion actually reduces to morality. Admittedly, a large proportion of religious practice has a moral dimension, yet does this entail the position that religion is purely a matter of listing a set of do's and don't's? The same is not automatically the case for religious experience, as one is not required to be moral in order to earn it. The apostle Paul, while still named Saul, is a glaring example of this. As Saul, he persecuted the church and saw himself as acting out the part of an upstanding Jewish citizen. The God of the text, however, did not see it this way and made Saul temporarily blind for his rashness. This occurred during Saul's religious experience of seeing a vision of the risen Messiah, and as such lends credence to the claim that morality is not a prerequisite policy for being in line for that kind of experience (not that Saul was actually seeking one). Thus, there does seem to be an aspect of religion that is not simply based on morality, not least of which is the very act of believing or having faith in the God to whom one is devoted.

(A discussion of the extent to which our behavior is freely chosen or controlled must await the later examination of behaviorism. In the meantime, we turn to the thought of Karl Marx.)

Marx

Karl Marx (1818–1883) found himself immersed in nineteenth century British liberalism as he set up camp in the Reading Room of the British Museum while in exile from his Prussian detractors. Liberalism was then understood as a child of Enlightenment thinking, where reason through education could propel humans toward and ultimately reach perfectability. Well before World War One was to demonstrate the unattainability of this ideal, Marx diagnosed liberalism as not truly liberating. He criticized liberal economics for its lack of equality. Capitalism was seen as liberating in that people were free to engage in commerce, but Marx maintained that the system was free only for those with the means to actually conduct business. While capitalism was and is defended as the most enlightened economic system and so should be promoted, Marx viewed it as "merely an expression of the self-interest of those fortunate enough to possess land, property, and capital" (p. 167). In a topic to be discussed more fully in Part Four, the system is actually conservative, not liberal.

Marx was a materialist and determinist when it comes to history: the former in that history encounters no outside interference from, say, God or Spirit (as it does for the philosopher G.W.F. Hegel), and the latter in that history assumes an unavoidable sequence. Marx sees history as marked by the rise and succession of economic systems, most recently feudalism giving way to capitalism, and in turn yielding ultimately to communism. For Marx, history is written by covering the themes of which economic systems surpass and supplant which. Yet he softens his stance by stating that history does not operate with scientific precision, though it would need to if it were deterministic. What is deterministic is reducible to mathematical description, and history, Marx admits, is not. This becomes an area of fragility in Marx's analysis, for he cannot have it both ways—if history is not an exact science, then neither can it be deterministic.

On the issue of human nature, Marx believes that there is no fixed one. Nor could there be, he recognizes, if it goes by social and economic descriptions. For if human nature is dependent on socioeconomic factors, then it will change in different times and places as the circumstances become altered. Human ills are to be found under all pre-communistic schemes, hence capitalism, as one of them, is ill-equipped to resolve them. By implication, the only solution to economic problems is another economic solution; meaning Marx's strategy for alleviating economic difficulties is to throw more economics at them. What we need is a good old-fashioned economic revolution, Marx would chime, where the endgame is communism. The trouble is that, contrary to his forecast, this revolution has not occurred in the advanced capitalist nations, such as in North America and Western Europe. What is worse, and also surprising had Marx witnessed it, is that some countries having attempted a form of communism can and have reverted to capitalism, the former Soviet Union being a case in point.

Perhaps the greatest drawback of Marx's work is his overestimation of human nature, for his belief that the remedy for our alienation and estrangement is a healthy dose of communism amounts to unguarded optimism. There is no guarantee that the "dictatorship of the proletariat" will assume a beneficent role once in power. The new form of government replacing the old may neither be classless, since those in power could opt to form a new ruling class, nor benevolent, as they might locate and devise "opportunities to abuse their power and develop new forms of exploitation" (p. 179). A different economic proposal, in my estimation, is not the solution, yet in quasi-religious fashion Marx views communism as salvific. Marx proffers a utopian secular faith that fails to deliver as he expects. The reason for this, to anticipate a later theme, is that wherever humans are involved, corruption will follow. Communism still involves people and as such will fall victim to distortion (Stevenson & Haberman, 163–79).

I wish to elaborate on this last point for a moment. In my opinion, Marx was correct in assessing communism as the highest form of socioeconomic and political policy to date. This form of government calls for all its citizens to conform to its ideals in order for it to work. It requires team players. This is not the natural inclination of humans, though. Where there is a possibility of corruption, some humans will eventually find it. Sadly, that would undermine Marx's vision. It may be the case that humans themselves would need to evolve morally to the point where a Marxist form of government could succeed. If this were to materialize, however, I suppose that a higher form of humanity might not even need to be governed. But this is not where humans currently sit. With their own ethical efforts, humans might not even be able to aspire to global just and fair dealings. Humans characteristically insist on their freedom, and should events turn sour and we experience, say, a total collapse of the financial system, I pray we find the strength to sacrifice our individual wills for the sake of the group. If not, our species might not last long. If our cry out of self-interest is, "long live the one," there might not long be a many.

Sartre

Our first look at a figure whose entire lifespan is within the twentieth century is Jean-Paul Sartre (1905–1980). Sartre is diametrically opposed to Plato in the sense of what the Forms imply. Plato understands there to be a realm of Forms before there ever were instantiations or representations of them in our world, since only Forms are eternal. Hence these Forms as essences precede the existence of their earthly counterparts, their pale imitations. Sartre, on the contrary, takes existence to precede essence, that is, we are the product of our own choices; we exist prior to constructing our own essence. That we exist first and then make our essence is based on the necessary exercise of our radical freedom. In this sense Sartre is also opposed to Freud, for our emotions do not determine us, rather we determine our choices, emotions being one of them. In tune with Kant, we are not caused. For instance, we are not forced to be sad, instead we elect to be sad. Sartre was also influenced by Marx, both by having a social conscience and acting on it in the form of pursuing justice for the exploited and oppressed. In Kantian terms, Sartre had the proper intent—his heart was in the right place, even if his or Marx's head might not have been.

Sartre's diagnosis of the human condition is such that we would prefer living in a Freudian situation where we have no choice about our actions and our behaviors are determined. This would release us from the responsibility of our choices, leaving us nonculpable for them since they are not our own. To the extent that we seek these circumstances, Sartre claims we are acting in bad faith, for we would do anything but admit that we were free to choose. We would much rather pass the buck of responsibility than accept blame and shame. We find ourselves in an absurd condition, for we are abandoned to act without making any legitimate appeal to an objective standard or transcendent code of conduct, especially since God is illusory. In the absence of divine commands, we are left to fend for ourselves. We are on our own and must make our own values, for no one and nothing can make them for us; and when we decide, we must realize that there is no external justification for them. We are "condemned to be free" and must choose how to create our own natures.

Our freedom is without limit and to recognize this is to be in anguish. We can neither escape our freedom nor the need to make self-conscious choices. We are forlorn and must face up to our freedom and affirm our choices; to do this is to act in good faith. And a steady diet of good faith produces an authentic life. A Sartrian credo or slogan then would ring something like this: being or existence by itself is nothingness, it must be filled with essence.

Here are some points of concern by way of commentary. Can any choice authentically made have value? Is exploitation as authentic as altruism? If value choices are unjustified, does that make them arbitrary? A response to these questions involves the respect due to everyone's freedom and the allowance for it to be exercised. Herein lies the difficulty. The trouble is that authenticities will clash. The choices and actions that make one person authentic might conflict with those of another. How then from a logistical and feasibility standpoint can everyone's freedom be respected, let alone exercised? A related issue arises for those who, in Sartre's appraisal, act under the illusion that there are in fact objective, even transcendent, values. Even were they to hold such a position, does this necessitate their acting in bad faith? Was Mother Teresa being inauthentic? More on this in a moment.

In his later writings, Sartre acknowledges that there are limitations to our freedom. Owing to Marx's influence on Sartre, there are socioeconomic factors that each of us must contend with, making this admission either a softening or a backtracking on Sartre's part. Hence there are restrictions on our freedom, for we are unavoidably affected by others and the rest of our environment, which in turn means that we are never completely free (Stevenson & Haberman, 184–99).

Sartre's ideals are noble and commendable. Individually we should seek self-improvement through the proper use of our freedom; and collectively we should enable all others to do the same. One might be prompted to inquire at this juncture as to whether this humanitarian spirit actually warrants the support of and commitment to an existentialist philosophical system? Or is it simply what many persons of varied philosophical and religious stripes would approve of and engage in? Lastly, we may further ask, and thereby, as promised, reintroduce a theme from a moment ago, as to whether God needs to be ousted from the system in order for us to be authentic? Can choices not be in good faith even were God the source of values? Would the presence of God prevent us from becoming authentic? The fact that there are (or were) those who call themselves theistic existentialists means that more people than expected can fit under Sartre's canopy.

Darwinian Theories

As a final installment of our distillation of topics found in Stevenson and Haberman's text, we turn to evolutionary themes subsequent to the work of Charles Darwin (1809–1882). Their selection covers Darwinian theories but is not confined to a treatment of Darwin. Those sections most germane to our purposes deal with the social implications (if legitimate they are) of Darwin's thought. The first can be accomplished in short order. If there is a Darwinian ethic that champions one value above the rest, it would be reproductive success. Success in biological terms is measured by the number of offspring left by an organism. Hence the value would be to leave more offspring, since that promotes the "good" of passing on more of one's genes to the next generation. If this is true, then has the Roman Catholic Church been unwittingly Darwinian all along? The Church's policy has been to counsel couples to increase the size of their "litter," thereby espousing an ethic designed to outbreed the pagans. One of the Church's hymns should then read, "What a friend we have in Darwin!"

The second social implication concerns a social conscience that would offset the otherwise cruel aspect of the survival of the fittest, a phrase coined not by Darwin but by Herbert Spencer (1820–1903). To this end, it is useful to quote the authors at length:

> Darwin voiced some worry about the multiplication of inferior types of people being "injurious to society," but he immediately went on so to say that we could not check our sympathy for the weakest members of our society "without deterioration in the noblest part of our nature." So the ethic of universal compassion and respect, or "love thy neighbor as thyself," should take precedence over any hardheaded biological calculations about theoretical benefits to society as a whole. Such arguments would be based on scientific hypotheses that might be far from certain. As Darwin put it, they "could only be for a contingent benefit," whereas there would be "a certain and great present evil" if we intentionally neglect the weak and helpless. (p. 210)

I would like to say that it goes without saying, but it actually needs to be stated, that "Darwin himself did not fully endorse 'social Darwinism'" (p. 208).

Having set the stage, we can expand on the issue of eugenics, the term originating in Darwin's own extended family, with his cousin Francis Galton bearing the responsibility for it. Inspired by Darwin's own misgivings about allowing the unfit (those physically and mentally diminished in capacity) to thrive and the effect this might have in undermining civilized societies, Galton set to work on at least theoretically manufacturing desirable offspring. Otherwise, as the reasoning went, social programs would not enable natural selection to take its natural course. Inferiority, naturally, is innate, and the sentiment arose that it must be eliminated or at minimum curtailed. Superiority, evidently, encompasses white European males. Thus the state should ensure its own genetic future by stepping in to practice selective breeding in the form of compulsory sterilization of "defectives." Perhaps surprisingly, the U.S. and Scandinavia, and not Nazi Germany, were pioneers in these efforts.

Difficulties with this sentiment include who is awarded the duty of evaluating people as infirm as well as of deciding membership in race. Research has clarified that racial designations can be "more sociological than biological," since complexion reveals a graduating scale rather than a "sharp distinction." To be a Victorian British male was to be both racist and sexist, Darwin himself not completely immune from such prejudicial classifications either. Even his position on intelligence was arrived at through avenues more intuitive and visceral than scientific. Herein he "fell below his usual high scientific standard" (p. 216) (Stevenson & Haberman, 207–16).

Behaviorism

Within their current chapter, the authors anticipate some contemporary discussions and we concentrate on one of them here, prior to a more extensive look in the upcoming Part Three. At issue is the ongoing debate as to whether nature or nurture has the most influence on our lives. On the issue of whether human behavior can be attributed more to genetic or environmental factors, behaviorist psychology initiates the first round.

Darwin devoted part of a chapter in *The origin of species* to the behavior of organisms, thereby implying that even these types of variations are heritable and either adaptive or not and have survival value or not. The ones that do are passed on to future generations, though he was unsure as to the physical mechanism undergirding natural selection—the process whereby useful variations are beneficial for the moment and manage to become expressed in another round of living.

Freud accepted this baton and ran with it. (As intimated, Freud, whom we have mentioned already, is covered by the authors in their previous volume but is omitted in their current one and replaced with a chapter on Buddhism. One must make way for the other and there is no room for both, since goodness knows, a text on *eleven* theories about human nature would be overdoing it. No doubt publishing policies enter into it.) Freud sought to place psychology on a sure footing by giving it a biological foundation. As we have seen, instincts play such a role for Freud, though later psychology was not to follow him in this. Whereas Freud was suspicious of our belief in the ability to control our own behavior, given that, for him, our repressed desires afford no such comfort zone, J.B. Watson, the pioneer of behaviorist thinking, agreed that behavior is determined but that Freud was looking in the wrong place for it. Watson's misgiving was directed toward anything mentalistic, that is, he did not regard mind, intention, consciousness and the like to be sufficiently empirical so as to be rigorously scientific. These concepts, as abstract rather than concrete by definition, could not be observed and so did not belong in any scientific account of humans. He urged that we should be looking to the environment as much more important than heredity in the shaping of behavior. For Watson, the environment is that which causes behavior, meaning if you control the environment, you control behavior. Behavior can be determined by manipulating the environment, so much so that Watson believed that "any healthy child could be trained or 'conditioned' to become" another Mozart, for instance (p. 221).

 B.F. Skinner then took the reins from Watson and argued that behaviorists were the true Darwinians in their account of human behavior. He pictured the environment as naturally selecting "behavior, rewarding or 'reinforcing' some behaviors so that they tend to be repeated" (p. 221). Skinner was convinced that the regularities of behavior could be reduced to law-like generalizations, or natural law, and he hoped to uncover these. To this end, he conducted experiments on rats and pigeons and held that his findings could be applied to humans as well. Whereas Freud supposed that the first five years are the most formative for human behaviors, Skinner assumed that it is futile to speak of innate factors of behavior, that all behavior can be traced to environmental influences, and that experimental animals like rats and pigeons can stand in for humans.

The pendulum next swung back in the opposite direction with ethologists such as Konrad Lorenz. Ethology is the study of animal behavior, and Lorenz maintained that innate factors were dismissed too hastily. Some behavioral patterns, such as those we call instincts, arise "spontaneously in all individuals of the species (or in all males or all females) almost independently of previous experience or learning" (p. 223). For Lorenz, some behaviors are inescapably fixed in a way that "cannot be eliminated" regardless of the extent to which the environment is altered. Some behaviors can be modified, but not the majority. Lorenz, though, was also not without his assumptions. He saw innate tendencies where instead there might have been cultural learning taking place; and he conjectured, often without testing, as to which selective pressures could have been operative in our ancestors that would give rise to these behaviors. In this way, Lorenz believes that ethology should be accorded greater weight than psychology, since the former reveals a closer alliance with evolutionary theory.

It must be stated by way of commentary that the last point sounds as though it bears a strong environmental component. In fact, aside from changes to the DNA molecule through mutations brought on by radiation, chemicals or recombination, are not most other selection pressures environmental? And does this not play into the hands of the behaviorists? And for comparison purposes, Skinner's approach can be understood as the opposite of Sartre's. While Sartre takes humans to pretend that they are not free, and thereby do not care to accept responsibility for their actions, Skinner sees humans as believing that they are free though not realizing that they are actually determined by their environment. Additionally, Skinner and Lorenz share the presupposition that inferences can be drawn from animal studies to humans, yet as we have seen they come to vastly different conclusions about the roles of heredity and the environment. Moral of the story: sameness of subject matter does not dictate sameness of conclusions.

At the risk of developing vertigo with these pendulum swings, there awaited a more balanced (and dare we say sober) voice, namely E.O. Wilson. Whereas Skinner was a Harvard psychologist, Wilson is a Harvard sociobiologist, who recognizes that neither nature nor nurture should be emphasized to the exclusion of the other. The proper view is rather to judge the degree to which each is a factor in a given situation (Stevenson & Haberman, 220–25). Given the pattern of the history of these debates, however, there could be further pendulum swings to come, so we had best stay tuned.

Part Three

Contemporary Material

As will have become abundantly clear by now, even painfully obvious, is the fact that there is no consensus on the human nature debate. The authors of the *Ten theories* text themselves opt for a combination approach, locating insights in each offering but electing Kant as a base and building a case upon that foundation (pp. 235–36). Our task at this stage is twofold: addressing first the theme of the mind-body problem and then the issue of personal identity. To this end, much of the material here is drawn from the introductory philosophical textbook *Twenty questions*, edited by Bowie, Michaels and Solomon.

When posing the question as to the composition of humans, two or three possibilities emerge, as already intimated. The first comes from materialists who announce that humans consist entirely of matter and all their constituents are physical. The Judeo-Christian scriptures afford us with two additional alternatives, though in so doing they reveal themselves as ambiguous on the topic. At a number of spots they seem to imply that humans are comprised of two classes of things: body and soul, the first material and the second immaterial. This approach is known as the dipartite view in that it dichotomizes the makeup of humans. But there is also a tripartite scheme in which the spiritual aspect of humans is either added on to or further subdivided into soul and spirit, thereby granting humans a trichotomy of parts. At times the Bible is unclear as to whether spirit and soul are functionally equivalent or whether a further separation is warranted. The passage in Genesis 2:7 informs us that we are made of the dust of the ground as well as the breath of life from God. When God breathed into our nostrils, we became living beings. This suggests that persons are animated bodies, that God as the bestower of life in essence booted up the system or turned on the ignition. Such a description need not imply that a separate entity or component has been added to humans that was not there before, other than, say, a push to start the system on its way. Keeping in mind, of course, that when it comes to including statements to the contrary, the scriptures rarely disappoint, for Ecclesiastes 12:7 indicates that at the point of death, an organism relinquishes this breath or spirit which then "returns to God who gave it." To the extent that this "it" is more than an abstract concept and actually bears what philosophers would call ontological status, a spirit would then be numerically distinct from a body. But this is to anticipate the treatment below.

Right from the outset, though, there appears to be a difficulty. Life begets life, which in itself is not scientifically problematic, but for those who insist that the divine contribution in Genesis 2:7 was in fact a soul, two questions arise. First, was there a stage in the evolutionary development of humans when a soul became a direct divine deposit into a creature that lacked it? Second, does God laboriously need to intervene in a similar manner each time there is a newborn arriving on the scene, or is soulness naturally passed onto and inherited by all subsequent generations? This has significant implications for genetics.

Mind-Body Problem

Descartes, widely touted as the first modern philosopher, held a dualistic position on this issue. Like Plato, he believed that humans contain a soul within a body and that each can exist separately from the other. Moreover, the soul is able to survive bodily death and is then carried into the afterlife. Unlike Plato, however, the soul is not eternal but had a beginning (its creation by God) and is subject to change. Since the time of Descartes, the term mind has been used interchangeably with soul. He took the mind to be numerically distinct from the body, that is, humans come with a body, a brain as part of it, plus a mind. This of course makes the assumption that the mind is connected to the body by way of the brain, though that need not be the case. Nothing requires the mind, if such there be, even to exist internally to the body but could surround it, or perhaps specifically the head, as a type of aura, to think of one possibility to consider. But we digress.

Dualists such as Descartes are faced with the onerous task of explaining how two entirely different categories of things are related. Physical entities like the body are nothing like immaterial things like the mind, so how can the two be connected? Ordinarily, we operate under Cartesian assumptions and, perhaps unwittingly, subscribe to the position formally known as *dualistic interactionism*, which holds that both physical and mental events are linked in a causal chain. If we, for example, suffer from at least temporary bad aim and instead of a hammer hitting the intended nail it lands on our thumb, this physical event causes a mental event, namely pain. There are those who would claim that pain and other subjective occurrences are all in the mind. Well that's the trouble. Pain hurts. Subjectively.

The next link in the causal chain is this mental event giving way to the response of another physical one, this time the bodily or verbal (or both) reaction of the utterance of colorful language or a choreography of contortions. Such is the sequence: from physical to mental to physical events once again. The main concern here is that these events have nothing in common to effect the alleged connection, no underlying substrate. Descartes envisioned that such a causal joint occurred in one's pineal gland, but this was later shown to be wide of the mark. We are thus left with the conundrum of what mechanism connects the two, as well as the aforementioned enigma of how the mind can exist in space if it is not spatial to begin with.

Our tendency is to assume that physical events can cause non-physical ones and vice-versa, that is until we reflect on it. In unguarded moments, however, the situation becomes surprising. We may scoff at the notion of telekinesis—the alleged ability on the part of some to move physical objects without any contact, simply through the power of the mind. Yet we claim to be doing a similar thing all the time. We decide to apply our fingers to computer keyboards or other electronic devices, though seldom are we conscious of the mystery here. We do not tend to ruminate on how, simply by willing it, one's arm can be made to move so as to effect the finger motion we intend. We are all given to render implicit assent to the parapsychological category of telekinesis in our own experience. The difference, should it amount to a significant one, is that our arms and fingers are internal to us, while the physical objects which those involved in telekinesis purport to move are external to them.

As a side note, *some* dualistic interactionists maintain that matter can exist without mind, as in the case of rocks for instance, but not the reverse. For this subgroup, mind cannot exist on its own, which differs from both Plato and Descartes, and which in turn has implications for post-death survival of the mind or soul. And on the subject of what science is competent to study, Descartes is one of the few major thinkers to hold that both sequences mentioned above are in play. The first is common to most if not all who deliberate on these themes: a) the body as something physical; b) is deterministic; c) and as such is subject to mechanical explanation, since it is machine-like; d) can be described using cause and effect; e) is reducible to mathematical equations; f) is public; g) and can therefore be studied by science. The second is defended only by a few: a) the mind as something immaterial; b) is the seat of free will; c) is not subject to mechanical explanation, since it is not machine-like; d) cannot be described using cause and effect; e) is not reducible to mathematical equations; f) is private; g) and cannot therefore be studied by science. Descartes' answer to the question as to what is the extent to which science can address all aspects of human nature would be that it can treat all but our mentality. The mind, for him, is forever beyond the scientific method to capture, try as it might.

According to Descartes, we would not mistake or confuse humans and machines (a topic to which we shall return) since machines lack the capacity to reason. Using Cartesian phraseology, humans are mind-brain amphibians, in that as amphibians are at home in both aqueous and terrestrial environments, humans are comfortable in both physical and mental domains. Employing another analogy, it is not as true to declare, as Descartes would, that the mind is in the body as a pilot is in a ship, for pilots can disembark ships when in port (or involuntarily while at sea), but the mind is not free to vacate the body unless the latter perishes, at least according to the rules of the Cartesian game.

One final item. Descartes agrees that the mind is not coextensive with the body or pervasive in it, for if it were, then mentality would be diminished for each and every body part that would be lost. We do not have the impression that we become deficient mentally if, say, we were to lose a finger. The person may be troubled and even go into shock with such a mishap, but the mind will still be present in its entirety.

Descartes' views continue to find fertile ground, at least in the hearts of his followers. Yet fear not those who are disgruntled, for other options abound. *Epiphenomenalism* is a position put forward by those disillusioned with dualistic interactionism. In contrast, epiphenomenalism urges that mental events are simply a by-product of physical events and thereby have no existence of their own. This alternative, introduced by Thomas Henry Huxley (1825–95), sees the line of causation as unidirectional, with physical events causing mental ones but not the reverse. The standard example of this is the casting of our shadow; it is a by-product of our presence but it does not exist on its own. Shadows do not physically influence the objects that cast them, not do they actually cast physical objects.

Should you find neither of these approaches adequate, then there are several other alternatives to consider. Another proposed solution to the mind-body problem is *physicalism*, which is the direct opposite of dualism. In physicalism, an example of which is referred to as *identity theory*, there is only one category of things (a doctrine known as monism), not two as in dualism. Only the physical exists and every occurrence must be interpreted in these terms. Materialists believe the same thing. Each mental event then can be traced to, or identified with, a purely physical event (and hence the name of this position). Mentality is therefore reducible to, or is "nothing but," the firing of nerve cells (neurons) in the brain. The desire to eat lunch, for instance, is neuronal, meaning it's all in the brain, as it were. There may be two different sets of descriptions, the physical and the mental, but they apply to the same event, and the mental set should be eliminated (the view of *eliminative materialism*). The trouble, however, is that there are organisms without neurons that still wish to have lunch, such as the amoeba, so there must be more to this situation than simply the firing of neurons.

Those giving identity theory a bad review, but wanting to remain in the physicalist camp, have opted for *functionalism*, where psychological terminology such as desire is not reducible to biological, specifically neurophysiological, vocabulary. They stop short, though, of claiming that there is anything substantial about mentality (or in philosophical language, that a move has been made from epistemology to ontology, thereby granting mentality the status of being). So what is the difference, then, between the two physicalist orientations? Functionally speaking, mental states do play causal roles. Yet how can the non-existent be a cause? Well, for example, an imagined predator can cause a fight or flight response. Elucidation will follow.

For now, we can place these positions on a spectrum for handy reference. On the extreme right is the viewpoint of idealism where only the mind exists. To the left of it resides Cartesian dualistic interactionism. On the extreme left sits the monistic physicalist camp in its multiple forms. In the middle lies the epiphenomenalist strategy, where there are minds only because there are bodies. In this instance, it depends on how the middle view is worded. In a more materialist and therefore monist posture, mentality has no reality, only physicality does. The other formulation seems to accept minds as by-products, in which these products are taken seriously as real. If, as is purported, minds are real only because the physical is real, then the wording appears more dualistic than perhaps is intended. The wording is crucial; a task for the philosophers if not also the attorneys (Bowie, Michaels & Solomon, 183–86).

Edwards

Philosopher Paul Edwards (see, we are invoking one already) takes issue with all three of these rival positions. Considering dualistic interactionism to begin with, Edwards discloses a concern with the standard sequence of events. The description physical-events-lead-to-mental-events-which-in-turn-give-way-again-to-physical he finds exceedingly glib, for it assumes that the transition from physical to mental and back again can overlook the unknown and regard it as ordinary or customary. On the one end, as is widely understood, intoxicants can cause hallucinations, and those we experience are known only to ourselves; no one else can experience our experiences. And on the other, the stress and anxiety we endure mentally can reduce our immunological defenses physically. Physical affects mental and mental influences physical. Explanations for this elude us. But it gets worse.

Edwards, in an idea not original to him, confirms that at each step the scientific law of conservation of energy is undermined. (This law in physics states that the total amount of energy in the universe is constant and is neither created nor destroyed, although it can be transformed.) The transition from physical to mental carries with it the discontinuity of a loss of energy, for it somehow disappears into immateriality, and then the discontinuity from mentality to physicality entails its miraculous reappearance. Hence there are two instances where the law is not upheld. The difficulty is that no evidence can be found supporting the notion that nature is ever thus duly interrupted. Losses in energy are not detected when sensations occur, nor are manifestations when volitions become expressed.

Edwards next has epiphenomenalism squarely in his sights. Recall that mental events here are understood as echoes of physical ones and nothing more. The only causes are bodily ones. Edwards agrees that the wording is important. As an initial response, echoes are physical in that they are termed longitudinal transverse waves. While waves by themselves are not purely physical, their substrates, like water, air and space, are, and waves are an accepted subject matter of physics. Is this what epiphenomenalists have in mind? Their reference to echoes does not make the point they intend, for sound waves are more physical than they expect and in the right circumstances can shatter glass. And in accordance with the aforementioned discussion on the extent to which epiphenomenalism can be considered materialistic, two approaches can be taken. In the narrower view of materialism, only physicality exists, making mentality a sub-category of the physical, if in fact the mental is also the real. In the broader version, material reality is the foundational one, but this does not exhaust all possible realities. Physicality may be primary, but mentality becomes a secondary or subordinate reality which lies beyond the physical. In Edwards' estimation, "somebody could be a materialist and at the same time allow that there are mental processes which are not a sub-class of physical occurrences. In this sense, dualism and materialism are not contradictory theories." Consequently, epiphenomenalists could find themselves in either camp. "In the broader sense, even quite a number of dualistic interactionists could be regarded as materialists" (Edwards, 81).

On one side of the ledger, epiphenomenalism allows for causal influence in the body to mind direction; on the other, this resolves only half of the difficulties plaguing dualistic interactionism. A reckoning of energy loss, as intimated, still needs to transpire. But there is yet another significant problem. If the assessment of epiphenomenalism is accurate, then all of our thoughts, beliefs and everything else that occupies our mentality are at the mercy of physical processes. Our brain and nervous system would control our subjectivity and we would be determined by them after all. Ultimately, not even the concept of epiphenomenalism would be based on rationality, evidence or logic. Certainty about anything would escape us, even (or except?) that our brains call all the shots.

Edwards further rejects a recourse into identity theory. Here the mind is purely an invention, and the language used to describe it conjures it up merely in the imagination but does not thereby give it an existence. (Note, however, that we must ask where this imagination would then reside.) Edwards agrees that mental events might accompany physical processes and vice-versa, yet this does not permit us to definitively conclude that mentality is physicality (Edwards, 78–82). That would amount to guilt or sameness by association.

Curt John Ducasse, who submitted the conservation of energy rebuttal put forward against dualism prior to Edwards, is similarly inhospitable to materialist views. The notion that matter is the length and breadth of all that exists and there is no reality beyond it becomes the source of his disquiet (Ducasse 1979, 87–91). He describes his misgivings in this way: "But a moment's reflection is enough to show that this conception of existence is perfectly arbitrary, as being not a hypothesis, capable of being tested and proved true or false, but merely a specification of the particular range of existence to which the sciences…choose to confine their interest. That conception of existence simply marks off their particular horizon…" (Ducasse 1979, 91).

There is an additional problem which buffets both epiphenomenalism and identity theory (a topic to which we shall return in Part Four). An example of mind affecting body is the placebo effect, where there is no biochemical justification for responding to treatment and improvement in health, since the administered medication is not the anticipated one, but may be as innocuous as a sugar pill. The point is that the patient believes the treatment to be effective on the authority of the physician prescribing it and trusts in the competence of the medical community. No rationale other than this belief and trust can be found for the efficacy of the charade. We thus find ourselves on the horns of a dilemma. On the one hand, physical science implies that minds do not exist based on the law of conservation of energy; and on the other, medical science implies that mentality is a reality based on this placebo effect. Scientific disciplines therefore militate against each other. Which should we be prompted to follow?

Two final items on alternative viewpoints. Some dualistic interactionists hold to mind as an emergent property of complex brain processes, though it enjoys no independent existence. God does not create a soul to deposit in a body, but the body awakens one on its own, for this might be the way God made complex bodies. Beside these two versions, and especially for those who are theistically inclined, lies the option of *parallelism* made famous by the philosopher G.W. Leibniz (1646–1716), under whose guidance the approach is termed *pre-established harmony*. Here body and mind exist independently and do not affect each other but run on parallel tracks which God ensures will proceed together in synch. The amount of divine involvement in this choreography can be extensive and means that free will is undermined, for God completely determines the co-ordination of the two tracks. Neither interacts with the other, leaving no causal connection between the two. Should the orchestration be entire and complete, it is given the name *occasionalism,* for every occurrence or occasion is foreordained. The parallelist position would be located on the right side of the two types of dualistic interactionism when it comes to the independent existence of body and mind, but to the left of the second version in terms of the effect of body and mind on each other. Figure 1 will help to situate the various positions along an axis stretching from mentality as real to imagined.

Figure 1. Mind-body/brain Relations

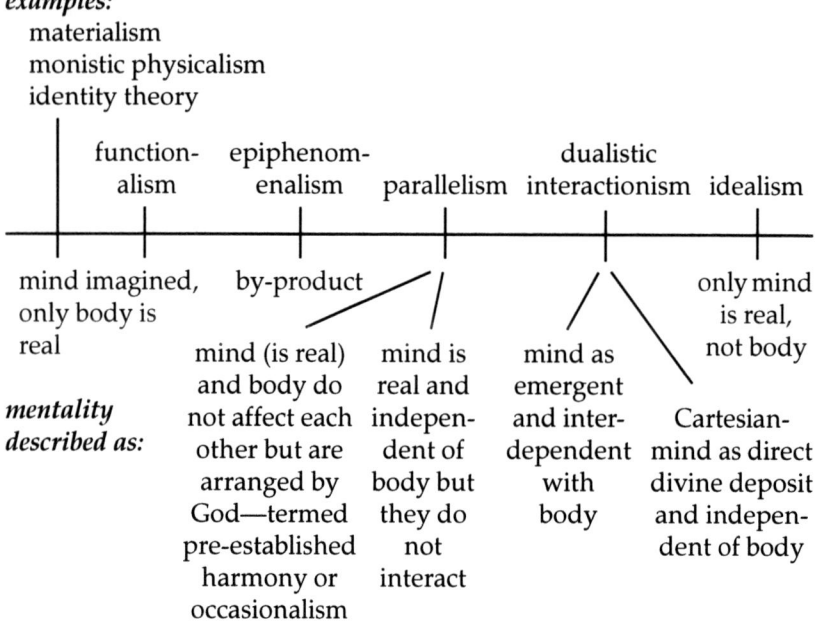

Ryle

Descartes comes under fire through the work of Gilbert Ryle (1900–76). The facile acceptance of mind-willing-and-body-executing is something that Ryle takes to task. The standard Cartesian model presents bodies as spatial and temporal (in time), while minds are temporal only—they take up no space, though he referred to both bodies and minds as substances. Material objects, such as bodies, influence other objects through contact, but minds do not operate the same way. If minds are not in space, then there is nothing to contact, hence causation does not mean the same thing here. Mental connections, should there be any, must occur through the mediumship of bodies. Ryle seeks to debunk this "official doctrine."

Despite our having direct access to our own interiority or subjectivity, no one else being privy to it, thinkers like Freud have informed us that even this access is not undistorted. To make matters worse, we are left to infer that similarities in the bodily actions of others is indicative of corresponding operations occurring in their own mentality as direct counterparts of our own. The difficulty we face is that body language is not as informative as we suspect, for it tells us nothing about other minds. In actuality, we are not even in a position to declare that other minds outside our own exist at all. Sadly, the evidence of bodily movements is insufficient both to surmise that a mind lies behind it, and if there is, that it experiences the same thing as we do when we respond with similar gesticulations. We expect that physical behavior leads us to the definitive conclusion that the agent is in possession of a mind which operates similarly to our own. The notion that a mind inhabits a body is a mistaken outlook that Ryle refers to as "the ghost in the machine." Since, according to Ryle, this is something we have invented, he labels it as absurd. Descartes understood mind as a thinking substance; Ryle views this as a contradiction, for substance is physical language and thought is insubstantial. As a proponent of the behaviorist school, about which we have already become acquainted, Ryle sees behavior as all the data we have to work with. This excludes mind and anything else unobservable (Ryle, 192–200).

Lycan

Next William Lycan believes that computers have something important to teach us about minds, so he addresses the topic of artificial intelligence (AI). Some definitions are in order before we begin. For starters, AI is "the science of getting machines to perform functions that normally require intelligence and judgment." Intelligence, in turn, is a "flexibility [or] responsiveness to contingencies," a type of spontaneity. Finally, a contingency is that which could have been otherwise or different from what obtained. As something conditional, it is the opposite of necessity. A contingency depends on the circumstances; should situations change, there are contingency plans in place for just such an eventuality. Lycan diagnoses humans as intelligent, for we demonstrate flexibility in "deal[ing] with the unforeseen" and modify our strategy accordingly. When the unexpected happens, we make the appropriate adjustments.

Lycan appraises computers as different from this. He envisions two limitations: first, on the input side, information must be introduced into the system, selected by "a human programmer or operator;" second, on the output end, a computer is incapable of interpreting data for meaning or relevance. In Lycan's estimation, a computer might have intelligence, but it lacks judgment. It has to be told what is relevant.

Lycan goes on to speculate as to whether computers are capable of duplicating human actions. If so, what would prevent us from conferring onto them a distinctly human property such as intention? And if computers were to gain mentality, then would we need to extend to them the basic human rights and privileges that we as mind-bearers enjoy, as well as the obligations we need to fulfil? We are thus asking what would need to be added to a computer for it to achieve the status of personhood? In so doing, we are engaging in the issue stated in the subtitle to this volume. We can also ask, from the opposite end, what could be subtracted from humans that would cause them ultimately to lose it? Let's consider the latter possibility. A cyborg is part human, part machine, examples of which include those with artificial limbs. A fyborg is one who is a functional cyborg, meaning supplemented with technological innovations. Most of us fit this description, at least those of us who elect to wear shoes, for instance. Now if a human were to undergo surgical operations to replace body parts piecemeal with mechanical ones, at what point would we be left with a machine instead? Would it occur once the majority of the brain becomes prosthetic, say 51%? (We will revisit a similar theme shortly. And kindly indulge the science/speculative fiction characteristics of several of these selections.) Or is there some indispensable part of the body without which a human would forfeit humanness?

Regardless of the extent of the cyborg operations, we tend to feel compelled to infer that what we have before us is still human enough to warrant retaining civil rights. Yet seen from the reverse angle, is it the case that the more complex the android, the more it is in line for receiving human rights? We discover, and this is Ryle's argument as well, that we are equally unable to determine the presence of mind on the basis of behavior. Lycan extends this discussion by stating that if machines are functional equivalents of humans, and this is a functionalist argument, then we become hard-pressed to make a successful, perhaps even legal, distinction. Behavior alone does not permit us to claim that minds lie behind it, something we commonly admit to for those organisms at the sub-human level. But if we are inclined to do so, then this leads Lycan to propose that, functionally speaking, minds are equivalent to computer programs. In functional terms, the two are indistinguishable (Lycan, 201–7).

Searle

John R. Searle then criticizes both of the foregoing positions by taking the previous discussion one step further. Contrary to Lycan, Searle does not evaluate computers as able to inform us about something important concerning the mind. Lycan's thesis, as we have seen, is that minds are very sophisticated computer programs, implying that computer states are mental states. To this brand of reasoning Searle applies the famous Turing test, devised in 1950 by British mathematician Alan Turing (1912–54), who conducted pioneering work in AI. Turing's test measures the extent to which one can be convinced that a system has mental states; if so, then it actually bears those states. It all depends on the success of the imitations or duplications. If we are confident that computer simulations are identical to the original, then they are emulations. If an expert in the field could not tell whether s/he is dealing with a human or a machine, then for him or her the system is virtually the real thing, the genuine article, and as such it passes the Turing test.

In language that Lycan would applaud, computers work with symbols but do not invest them with any connotation. Symbols for a machine are there to be manipulated and carry no other ramification beyond this. Searle interjects, however, that manipulation of symbols is not the same as mind. He employs an analogy to drive his point home. If we were to be given an instruction booklet on how to arrange Chinese characters in such a way as to make sense of the language for a native Chinese recipient of the message, then after achieving a certain level of competence, the Chinese individual could be satisfied that a person with facility in the language was sending the message. This would suitably meet the requirement of the Turing test. Yet all the while, not a single word of Chinese might be understood by the sender of the message. This sender is merely informed that if specific symbols appear in a particular order, then they will make sense to a native reader.

What this analogy reveals for Searle is that syntax—knowledge of the appropriate order of symbols—does not by itself lead to semantics—knowledge of the meaning or content of the message. Arrangement does not guarantee infallible interpretation. The upshot of all this is that these abilities can be faked, though still able to pass the Turing test. In Searle's appraisal, the Turing test is thereby refuted, for this example demonstrates that we could pass the Turing test without having the requisite mental states for it, which was the issue from the outset. We would not in fact possess the corresponding mental states of familiarity with Chinese, hence the Turing test is not foolproof. Contrary to Ryle, the Turing test "is simple-minded behaviorism." And contrary to Lycan, "computer programs by themselves are never sufficient to produce mental states." In essence, brains produce minds, programs do not, and we need more than running a program to reproduce mental capacities. The real mistake, Searle concludes, is to suppose that simulation is duplication, that a likeness or imitation is the real thing. Nor do we regard them as the same, for this is why consumers pay less for knock-off products (though these brands are not likely to fool the experts)(Searle, 207–13).

By way of comment, in all fairness to the Turing test, it rests on our (in)ability to make the assessment as to whether or not something is a fabrication, not on the actual state of affairs. And this is Ryle's point: the power to recognize the difference between behavior and the very presence of minds (which applies to both Ryle and Lycan). The first depends on appearance, the second on reality. We may be no closer to the second after ruling on the first. This brings us to an illustration from the theatrical/cinematic world. Actors, themselves students of human nature, convey by external signs what we are made to believe is occurring with them internally, though they might not actually be experiencing this at all. The best among them are very convincing in their craft and are rewarded for their efforts. If we cannot tell what is going on inside actors through the roles they play, then how much more difficult, we recognize, would it be to gauge the feelings of a machine, if AI were ever to get that far? As Lycan admits, there is no way of telling what silicon circuitry would feel like to its possessor, so we should not be quick to make a judgment about it. Perhaps even robots will enjoy interiority someday (Lycan, 206–7). Yet the same problem manifests itself when actors get in character. Could Robert Downey, Jr. really be British?

If Searle's diagnosis is correct, then machines could not attain subjectivity, for to do so would be to achieve personhood. At most, they could manufacture the outward appearance of it, but this would not amount to an improvement over Ryle or Lycan. Consequently, AI too has its limitations. We do not necessarily get any further ahead in understanding human nature by including computers in the mix, for they are only partially helpful in informing us about brains and minds. As a result, we can continue to ask if there is something characteristically human which a machine could not be expected to duplicate. Perhaps they will forever be unable to reflect on the kind of philosophical questions we are posing. What would a computer perspective entail? What would be their considered opinion on matters of taste? Jazz or blues? And would they be as curious as we are about whether we are alone in the universe?

Personal Identity

Discussions as to whether or not to call robots persons leads us to switch our sights to an examination of personal identity. At issue here is what accounts for our notion that we are persons and that our personhood persists from birth to death, and perhaps even beyond. Among all the changes that we undergo in a lifetime, what makes us think that we are the same persons from infancy to elderly? A number of approaches can be taken on this score. The first concerns the idea that personhood is somehow tied to our mentality, for we carry with us our personalities from moment to moment, and this, we feel, provides the anchor for what we call us. This view comes in two versions. Descartes, for one, would be of the mind that our soul makes us the same persons who performed some act long ago. This, of course, has implications for the legal system which rules that, for instance, there is no statute of limitations for crimes such as murder. Some courts have decided that, say, Nazi war criminals should continue to be hunted down and brought to justice even though their acts were carried out in the distant past. Are they the same persons now as they were then? These courts think so. Additionally, a benefit derived from the soul strategy is the conception of survival, that is, since the soul is immortal, there must be some existence for it in the hereafter, a life after life-and-death. The second version is the memory theory, where our recollection of our having perpetrated an act connects us to the person who carried it out, for no one else could have our very memory. John Locke champions this position.

Another perspective takes the body as the link from one stage of life to the next. Despite the numerous changes that bodies undergo during a lifetime, so the reasoning proceeds, the basic pattern of the body provides the continuity that we seek. This somatic proposal, as it is called, also has a second approach. The brain as a specific part of the body is sometimes seen as the most important organ and even the seat of personhood, so considerable attention is devoted to it in particular. Lastly, there are those who argue that personal identity is simply an illusion. For them, the self is a fiction that we have concocted from the myriad individual and separate experiences we have encountered throughout our lives. In order to connect the dots, we have invented a unity that we refer to as "me." David Hume is of the opinion that we do not actually persist through time (Bowie, Michaels & Solomon, 329–31).

To recap, mentalistic theories come in the form of 1a) soul, and 1b) mind or memory perspectives; somatic theories arise in the shape of 2a) body, and 2b) brain approaches; and Humean along with Buddhist analyses strenuously declare "none of the above."

Perry

On occasion, one way to place differing positions in greatest relief is to initiate a debate between the two. Plato recognized the merit in this and left us his writings in the form of dialogues to great effect. We begin, then, with two views squaring off—a showdown between the soul and body theories (1a versus 2a). There are not many contemporary thinkers willing to adopt a Cartesian stance, so a character most likely to accept such a role would be a clerical figure. In this case, John Perry, the author, opts for a hospital chaplain named Sam. His interlocutor is a dying patient named Gretchen, a professor of philosophy. Some of the highlights of their interaction follow.

Though she believes neither in a soul nor an afterlife, Gretchen wishes to receive comfort from Sam. She encourages him to present his case and allows for the possibility, but not probability, of continued existence. Sam launches into the diatribe by contending that persons will survive death and live again, whether or not the latter includes a body. The one who experiences now in this life is the same one who will experience again in the next. Gretchen responds by stating her concern that there is no known way to identify the new person as having been the previous old one, other than in empirical (observational) bodily terms. She asks what certainty there is that the new person will actually be me? Her difficulty rests with the concept of resurrection and what this will yield. Sam claims that the new person will be a duplicate of the old, "exactly similar" to us here in the now. Gretchen, however, reacts by insisting that "exactly similar" is merely wordplay. She holds that the issue is not exactitude but identity, the new person needs to be the very same as me.

Sam counters with the belief that the soul or mind affords this possibility and that we will be as conscious then as we are now. But

Gretchen asks, "What is the subject of the verb" conscious? Will it be the same me that is conscious? And what can be conscious but bodies? If our spiritual side constitutes us, as that which makes the crossing from here in this life to there in the next, then we are not actually the bodies in which we currently find ourselves. Yet Sam is adamant that the soul is what survives, making it the very person in question. If our souls are identical to us, then Gretchen reasons that we cannot be our bodies, for we, Sam included, use corporeal identifiers when recognizing a person. How do we know a person retains his or her identity from one moment to the next? Sam argues on the basis of "same body, same self." We are "intimately related" to our bodies "but not identical" to them. We perceive the same body and assume "that the same soul is connected" to it. Herein lies the problem for Gretchen. Our present bodies will "rot away," so we will have different ones in the alleged afterlife that Sam accepts. By his logic, "different body, [must mean] different person."

Sam is left to infer that, based on empirical evidence, the same outside must contain the same inside. Gretchen argues that there is no way to test this hypothesis, since the soul is non-empirical. Sam interjects with the use of personality as an "intermediate link." A person's traits are observable and we can correlate, even identify, them with a person's corresponding soul. Gretchen declares that "similarity of states of mind" is no guarantee that the soul is the same. Bodies can change and so too can psychological characteristics, implying that this does not necessarily connect us with the person "underneath." Since psyches are beyond the senses, how would we know if one had not been substituted for another similar one? Sam announces that he knows it to be true in his case, for his body and soul are "consistently found together," to which Gretchen retorts that even if this were established, it could not be universalized to all those now living. Buddhists, for example, would consider this contrary to their own self-perception. Moreover, she doubts even the capacity to establish it, for we have no way of knowing if same person means same soul, as in the case of split personalities. For all we know, souls can come and go, and we would need to be able to identify one first so as to gauge its sameness or difference, and this is still not forthcoming (Perry, 332–48).

Souls for Gretchen, then, fail to provide the mark of personal identity and cannot be employed as the basis for our continuity from current to subsequent lifetimes. The discontinuity of our bodies prevents this, making souls inadequate indicators of survival. As a comment, it might perhaps have been strategic for Sam to assert that despite our inability to determine the sameness of person or psyche, the issue would not be our own power but God's to make the distinction. God, after all, is the one allegedly orchestrating resurrection and has probably worked out the logistics beforehand so as to meet the challenge. Plus, God would not be reliant on the empirical world in making the assessment. Whether or not we have any emotional investment in Sam's position, we may wonder why he did not play this card.

Locke

Such, for starters, is the look of the terrain. The next stop on this personal identity tour is the second mentalist alternative, namely memory theory (1b). The classic statement of this perspective is submitted by John Locke (1632–1704). A distinction to make at the outset is the significance of mind in containing memory stores, but consciousness as that which recalls them—a retrieval system of sorts. Locke considers personal identity to extend as far in the past as we are conscious of our earliest memory. Reflection on a previous act constitutes sameness of person. Consciousness provides the glue that forges a continuity from earlier performing self to later reflecting self. As mentioned, no one else is privy to our thoughts, memories and experiences, nor are we to theirs. We are all unique in this. If we recall the day we got married as something that we formerly experienced, then our present recollection of it makes us the same person who actually had the experience (Bowie, Michaels & Solomon, 330).

Having stated the basic position, several comments are in order, some of which Locke anticipates. First, no one boasts total recall. We neither can recall a meal we had one thousand days ago, nor obviously can we, according to Locke's stipulation, retrieve anything prior to our oldest recollection. What's the first thing we remember? Well we do not recall anything before that. Hardly a controversial claim. We usually do not have a vivid memory of our umbilical cord being cut. Does this mean our initial forays into life were not with the status of personhood?

Second, even the ablest memories can err. We can be subject to faulty memory, for we can falsely believe that we left our keys on the counter when in fact they were in our pocket all along. Silly us. We suffer from inaccuracies of memory, especially with increasing age, as well as from imprecision in details. Does this mean we are not the same persons as those with these mistaken experiences? For such a "memory was merely apparent, not genuine," and the requirement is that we both have the experience and recall it at a later date. Failure on either end is detrimental to the theory.

The third and final comment amounts to an illustration of the above. Those who suffer from Alzheimer's disease might have had experiences for which they can no longer form memories. Some experiences may no longer be recalled, while others may contain faulty details. Family members of these patients might find themselves asking the pertinent question, can this really be the same person we knew before? On Locke's account it would not. If memory is the tie that binds us into a continuous self, then it is a loose one. Compounding the problem are such occurrences as changes in character qualities, sometimes prompting a spouse to exclaim in despair, "that's not the same person I married."

Hence there can be discontinuities to our consciousness as with the interruptions due to memory loss, blacking out and even sleeping states, though consciousness is not strictly lost in the latter. In sleep, we are not totally oblivious to our surroundings and still manage (or have the presence of mind?) to swat away an insect and remain asleep.

Enter a discussion of a different sort. Locke, having Christian sensibilities, extends the argument into the domain of the afterlife. Reinserting the same distinction between soul and consciousness, Locke reasons that the survival of the soul by itself is insufficient to connect the same person from one life to the next. It must be accompanied by the consciousness, for the latter carries with it the memories of the earthly life into the new. Else the judgment of God would be incomprehensible if a person had no memory of good or ill deeds for which s/he is being held accountable. Further, this is the mode of how a person will be recognized in the hereafter, for, contrary to Gretchen, the one who encapsulates the memories of your mother will be your mother. Consciousness unites these two lives into the same person.

Herein marks the difference between reincarnation and resurrection. In the former, consciousness does not carry over from one life to another, meaning there is no memory of previous existences. This, in Locke's view, would be enough to declare that with reincarnation there is a new person with every life. The opposite is true for resurrection, for in this case there are but two lives and memory crosses over from one to the other, ensuring that the same person is preserved throughout. The same soul or mind "without the same consciousness" does not "mak[e] the same person" (Locke, 349–54).

Leiber

A work of fiction that investigates the question of personal identity from the perspective of memory (1b) is Justin Leiber's novel *Beyond rejection*. This is a futuristic tale where bodies can be frozen and minds stored. In the event that there is irreparable damage due to bodily injury, the medical system recommends that everyone place their minds on tape on at least an annual basis, if not even more frequently, for safe keeping. The physicians believe that the length and breadth of personal identity is stored on those tapes, whatever it is that makes us us and not someone else. Taped minds can later be retrieved from storage and implanted onto blank bodies. Sometimes minds are taped and stored in situations where bodies are frozen until such time as science develops the technology to treat a presently incurable malady, that is, when bodies are more likely to recover under what will then be currently existing techniques.

Leiber foresees three difficulties with this procedure. The first is the physical one of obtaining a compatible match of mind and body, particularly if there is an appreciable gap in ages between the two. The connection of the two might not "take." As a remedy, there are what have been termed somaticians—health professionals whose task it is to see that the graft is successful. The second is like the first: there is the psychological shock that stems from the mind's foreign environs, especially if the new body is of a different physical type, such as opposite gender or different race. For this specific trauma there are psychetricians. And the third also proceeds from the first: should there be a large gap in, say, youthful mind and aged body, the new person will be out of date. For a mind that was taped perhaps a generation ago, all that was familiar to it may now be gone in the new world. Consider it the "Rip van Winkle" effect. Friends and family may have disappeared, and the skill set mastered in the distant past may no longer be relevant. With the rate at which technology progresses, the new person might easily be a technological dinosaur. To tackle the first two problems, scientists have developed what is known as a harmonizer to alleviate some difficulties, yet there is little assistance with the third. Retraining would be a monumental chore. As it stands, the success rate of harmonizer-aided matches is 80% under ideal conditions for the first attempt. It drops to a mere 5% for the second. The new world that new matches wake up to is too much for some.

One might ask, "And just who is going to pay for all this?" Well the government underwrites one's initial implantation, though it might not be of high quality. Recipients may need to wait up to two years for the privilege of obtaining a remainder, since only the best bodies go to those who can afford them. Therapy to assist the fine-tuning of matches is also government-subsidized. The entire surgical-therapeutic process can take months, assuming the two can adjust to the new arrangement.

Leiber then makes further stipulations. The rules of his game include the eventuality that minds have an expiration, or best before, date. There is a maximum of about a century of experience before the onset of senility, regardless of the running-time of the body. While in storage the mind does not age, because it does not experience. The clock is ticking only when minds inhabit brains, for then they experience. Thus minds over 85 years of age are not allowed to be implanted due to the prospect of insufficient remaining longevity.

Hence in Leiber's estimation, the mind with its store of memory and experience constitutes the person. He employs a computer analogy to illustrate his point. He likens the brain to computer hardware right out of the box, where the hardware is like a blank human brain in which no program has yet been loaded. The software then stands for the mind. This is what Locke refers to as a human at birth—a *tabula rasa*, or blank slate/tablet. As an empiricist, who argues that knowledge is gained through experience, as opposed to a rationalist, who contends that it is gained through reason, John Locke, whose thinking we have already encountered, understood humans as having no preconceived notions or innate ideas. That is, we come from the factory without knowledge and must acquire it along the way. Newborns, then, are like blank brains since they have not had experiences to date but must learn everything they know. On Leiber's account, a mismatch of mind implanted into what it perceives as a foreign body elicits a violent response on the part of the software against its new hardware home and consequently rejects it (Leiber, 357–62).

Leiber's work prompts the following questions. First, does he have the proper grasp of personhood, such that were the technology available and the procedure achievable, then minds and therefore persons would be transferable? This warrants a second concern. On the subject of transferring minds, how different would this be from resurrection where the soul and its memory allegedly cross over from the previous life? To recap from the section on Locke, reincarnation refers to consciousness and its memory recall as not carrying over from one life to the next. The body is new and the mind or soul is old, but there is no recollection of a past life or lives. In resurrection, on the contrary, there is a new and, this time, imperishable body as well as a crossing over of consciousness, meaning a recollection of one's lifetime experiences. (Some might be inclined to wonder if there are multiple pre-resurrection lives, where reincarnation and resurrection work in tandem, and perhaps even where previous lives can be accessed through regression hypnotherapy.) A third category in these proceedings would be resuscitation, where a person has become comatose or died but has been revived, leaving the individual to endure another death. The question then becomes which of these three is Leiber's view most akin to? As with reincarnation, in Leiber's tale there is a new body and the old mind, but unlike reincarnation the mind comes with consciousness and thus a recollection of past (or original) implantations. As with resurrection, the consciousness crosses over, though in resurrection the body no longer perishes. In resuscitation, the old body remains united to the old consciousness, though in Leiber's case there is an old consciousness in a new body. His scenario makes it dissimilar to all three, meaning the need may arise for a fourth category.

Michaels

Another study involving a comparison of two positions, though without a debate format, comes to us by way of Meredith W. Michaels. The contestants are the two ends of the somatic theory, and the question becomes what is more likely to carry personal identity, the body or its brain? In her fictionalized account, Michaels presents two college students, let's call them Jane and Joe, who both meet with a fatal mishap. Jane's body, though not her head, is crushed by a steamroller, and Joe, as a witness to this, suffers a brain stroke. A transplant is later performed, or perhaps more precisely a salvage operation, and Jane's useful head is attached to Joe's otherwise useful body. The resulting issue, then, is who would be the remaining person? Which takes priority, the body or the brain? Is this Jane by virtue of the brain or Joe because of the body? And in the most practical terms, which set of parents is held to pay for the resultant student's tuition?

Michaels mentions that the temptation is to assume that wherever our brain goes, we go, meaning the residual person would probably have the conscious awareness of being Jane, since the brain was her contribution. After all, on Locke's thesis, she would come equipped with all the memories of Jane's life experiences. The point that Michaels wishes to make is that memories come with reference to the very body with which the original experience occurred. In the example of learning to ride a bike, brains alone are insufficient, as bodies very much come along for the ride and even contribute muscle memory. So in this case, would the resultant student be the one who performed the cycling with its body or the one who remembers having done so with its brain? And to make matters worse, what if one of the contributors never learned this skill?

Michaels suggests that Aristotle's views ought to be revisited here. His position is that identification of self is done through the body. Michaels introduces the following example in order to reinforce her argument. I modify it here and apologize in advance for the gruesome nature of the illustration. If you were informed that you were about to be tortured, but that mercifully your brain was going to be removed and replaced by another prior to the event, would you still feel anxiety over what your body is about to endure? You will not suffer physical pain from the event, though the pain could be psychological instead. Is there reason to fear any sensation of pain if your brain is elsewhere? Is the person who undergoes torture the old body (the one vacated by the original brain) or the new brain?

Michaels hopes to impress upon us the notion that the body is an important factor in considerations of self-identity, and as such should be regarded as significant for who we are as persons (Michaels, 354–56). The lone comment I wish to make concerns the idea of different brain, different person, for this brings us to the topic of Alzheimer's disease. In these cases, the brains are structurally though not functionally the same. Hence could it also be said that a person can be different even with the *same* brain?

Appiah

A final instalment of the somatic theory is submitted by Anthony Appiah, who elects to avoid discussions of the brain altogether. He finds that no reference to it is needed when other issues such as race and gender are involved. We can survey his approach this way. He asks to what extent we can undergo physical changes and remain intact; or at which stage would we be inclined to wonder, "but would that still be me?" If we were to alter our hair color, for starters, most could agree that we would still be the same person. If we subjected ourselves to a modification of any number of physical aspects so as to determine if we are still us, how successful would we be? If we were to lighten our skin color, the likes of which Michael Jackson underwent, then would we still be us? Was Mike still Mike? Though he appeared different, we would not likely confuse Michael for someone else.

This prompts the question as to whether there is anything that we can do to change our form in order to change us, so that we would become different persons? Appiah considers race and sex as potential candidates. We will address the latter. If we were to submit to a sex-change operation, that is, gender-reassignment surgery, would that make us different people? There are those who have sought this procedure and later reported that the new gender is "the real me, the one I have always really been all along." So for them, they are now a different "me." Others might disagree. Note that biological sex status is male or female, whereas "gender [is] masculine or feminine, the social roles" (Appiah, 373) (bypassing for our purposes bi- or transsexual alternatives).

While personal identity might be ambiguous in this example, consider another. If we were to have been born the opposite sex (sidestepping those other than the two standard statistically most prevalent ones, such as hermaphrodites), would we be different people? Significantly, we would bear a different genetic makeup. Some will be compelled to assert that being born the other sex means a different person. Unambiguously for them, in this instance, the "me" has been lost. Does this imply that bodily changes that occur to us only subsequent to birth are the ones that leave us the same person, but in the womb is where this person-formation takes place?

Appiah interacts with the belief that the real essence of a person is the set of chromosomes people inherit from their parents. Were this assessment correct, then we would have a slightly different set if born the opposite sex, meaning we would in fact be a different person. Appiah also points out that a common misperception surrounds racial identities which are evaluated as less "conceptually central to who one is than gender...identities" (Appiah, 375), and prompts us to appreciate the importance of race. The following comments about gender appear to be in order.

We now boast the technology to alter our chromosomal makeup. These come in two types. Somatic gene therapy involves a modification of DNA in non-reproductive cells. This "affects only the current individual" and "the effects...may not last." The change "may not be long-lived, so the therapy has to be repeated." This procedure amounts to "a quick fix." Germ-line therapy, on the contrary, modifies reproductive cell DNA, hence the effects will extend to "every cell in the future adult organism," as well as to generations beyond. The change that is passed on will thus naturally be repeated.

Here are some implications. First, if we are the set of our chromosomes, then we become different persons each time genetic therapy is applied. Second, identical twins carry the same genetic complement, so according to this logic, they must be the same person. Nor have we even begun to treat the serious unrelated issue of whether or not comatose patients have lost their personhood. If people are unresponsive, are they still persons? (Appiah, 372–76; Gazzaniga, 375–77).

The "I's" have it

While we are on the topic of personal identity, a useful exercise to undertake is one which yours truly also participated in as an undergraduate student in Oriental philosophy class. The question which the professor posed is one which Hinduism tackles, namely "what is the I that has a body?" We use the personal pronoun constantly, but what do we really mean by it? Hence the quest to locate the elusive "I."

We begin, as did "I" those many years ago, by considering whether the I is in fact the body. Employing similar reasoning as for a previous example, if we were to lose a body part, say a finger, would this have a lasting effect on our I? Were the I coextensive with the body, then to diminish the body would be to diminish the I. Yet this is not our experience. We do not have any less of an I for having lost a finger. The personal pronoun still applies as much after as before.

What next? Could the I then be understood as the brain? To undergo cerebral trauma is potentially to yield different character qualities, depending on the nature and extent of the damage. But the point is not how different we become but whether we still carry with us the concept of an I. That we can respond in the affirmative is evidenced by the self-recognition on the part of the patient who might inquire, "What am I doing here?" Extending the discussion, we can further ask if the I is the mind. We might sometimes have been overheard saying "I wasn't thinking," or "What was I thinking?," or even "I lost my mind." If the mind is something that the I can lose, then the I must lie behind the mind and reside at a more fundamental level.

Is the I consciousness then? Well a similar response obtains here in the statement "I lost consciousness." If we can suffer (for we would not actually experience it) a momentary loss of consciousness, then the I must also lie behind it at a more fundamental level. Amplifying on the distinction to be drawn, in two states where consciousness is thought to be absent, after we awaken from sleep we tend to recall some dreams, at least for a while, yet we usually have no recollection of what transpired during a period when we are under general anaesthetic. What then is the case with self-awareness? We can say that we are aware and that it is something we also can have. Thus, in what is becoming a standard refrain, the I must lie behind it at a more foundational level.

How about personhood? We normally say that a person is someone we are, more so than personhood is something we possess. In either case, the I lies underneath. Finally, how does selfhood fare? Utterances here could include "It's me," or "I gotta be me," and especially "I was not myself today," as well as "That's not like you." If me is that which the I has gotta be, then the I lies still deeper. In actuality, the level of the I is the most fundamental and this is what Hindus are attempting to impress upon us about our atman. Much meditation is required to arrive at this ultimate self and several layers need to be peeled back before we can get there. We begin to appreciate the enduring problem.

Hume revisited

In an effort to be thorough, we must complete the circle and allow Hume to have his say. Like the Buddhists, Hume asserts that there really is no self as such, and he comes to this conclusion using the following rationale. He admits that we have a sense of selfhood, but in what "sense" does this "sense" consist? For Hume, it stems from our impressions. Personal identity implies continuity, but when we look to impressions to ground it, we do not find any. Impressions are many and varied and "are all disconnected." Where we want connection we imagine it; and when we want constancy we construct it. Our character qualities as well as our self-perception do not persist throughout a lifetime. So the problem is that we tend to refer to that which changes when attempting to find a basis for that which is constant. We point to that which is fleeting in order to situate sameness. This strategy fails, for discontinuity is injurious to identity. Impressions, therefore, are inadequate indicators of the self.

In Hume's appraisal, people "are nothing but a bundle or collection of different perceptions, which succeed each other and are constantly in flux." This does not offer a foundation for a continuous self. So what makes us suppose that we possess a smooth, uninterrupted personhood from infancy to elderly? What prompts us to invent a tie that binds all these disparate impressions and perceptions together into a unity? Do we simply make the association where there is none to be made? Contrary to Locke, "memory does not discover identity" but manufactures it. Impressions are merely isolated dots that remain separate, and we make the connections—we connect the dots. Hence the self is a fiction that we have fabricated (Hume, 365–68).

A compelling argument, though not entirely satisfying for some.

MacIntyre

There is at least one author whose vision of fiction is appropriate for human nature. Alasdair MacIntyre sees humans as raconteurs, story-tellers. If asked to furnish an autobiography, we offer a narrative about our view of how we arrived at where we currently find ourselves. Such a narrative might not appear coherent when faced with a crisis situation or a turning point in our lives, for we may have momentarily lost a sense of its trajectory. But we tend to craft one when the dust settles. In a linear fashion that we have inherited from the Judeo-Christian tradition, originally an oriental perspective adopted by occidental (Western) civilization, our self-perception and the accounts we give of it is in the form of our having come from somewhere and are leading elsewhere. Our tale follows the arrow of time, thus making sense of our lives as so many steps all geared to some destination or series of them. Since the arc of our lives has not yet extended beyond the present, the story of our future lives is a fiction and we present it as though we are characters in a novel.

The dramas in which we place ourselves render unto us the personal identities we may seek by offering us the continuity of fictional characters. For those who fail to secure such unity, the current of their lives seems or has become unintelligible. During these times there appears to be no value distinction between opting for one path as opposed to any other. Their compass has gone missing, and if they can disclose no aim, goal or purpose to their lives, no future end point, then their ability to launch out on a certain course of action might become impaired if not crippled. We locate meaning in our stories as part of a network of stories. Narratives become modified during the course of our lives and then get enacted, characters that we are (MacIntyre, 368–71).

The only comment I have about this otherwise healthy and life-affirming outlook concerns the modification of our stories. If at one point we declare that our lives are directed by the preparation for becoming and then the performance of being, say, a physician, but later discover instead a more appealing path followed by embarking on it and going through the exercise of how our lives now resume intelligibility, how then can we claim that our personal identity has not encountered discontinuity? Where there was once a telos (purpose), what Aristotle (whose work MacIntyre applauds) would call a final cause, has now undergone a break or rupture and has been replaced by another. Can such a derailment really be understood as exemplifying the continuity that a personal identity requires? Perhaps if we come to an agreement with ourselves that all previous accounts are to be regarded as null and void, can we be confident that the new one now makes sense of ourselves where another could not. Yet how many rewrites can our lives endure? Are our characters perpetually malleable? Or does there arrive a point when the most recent effort to account for our lives strains credulity and can no longer be considered robust? And have the psychologists reserved a term for this?

Progress report?

The time has come to take stock of the ground we have covered thus far. Has our path brought us any closer to knowing who we are?

Here is what we do not know. We don't know how many parts comprise us—one, two or three: body, mind/soul and spirit, or some combination thereof. We don't know if we have a mind beyond the brain, nor how many components a mind itself might come with. Neither do we know definitively whether persons are more influenced by nature or nurture.

We don't know what an I is or even if we possess one. Nor do we know if computers will ever boast them. We don't know if we have a personal identity and, if so, how it endures through a lifetime and perhaps beyond. We don't know if our soul, if we contain one, pre-existed our bodies and, if so, where that might have been. Nor do we know specifically what our fate might be in some type of hereafter. And should there be a form of afterlife, we cannot be certain that the identity we have now, if any, will be identical to the one we would enjoy then.

Hence we know very little. In essence, we do not know the us that we are. It seems that human nature is difficult to pin down.

Humans are not so easy to capture scientifically or philosophically. Typical humans.

So *quo vadis?*, where do we go from here? Can we make any headway?

PART FOUR

Recent Developments

Anthropological evidence

How shall we proceed? Perhaps we should start from the beginning, when humans first appeared on the scene. Yet the question as to when humans emerged from an infra-human species is difficult to determine and depends on one's definition. The issue as to what constitutes humanity must be addressed first in order to ponder when it occurred. The discipline of anthropology assists us in teasing out the details, but we are left with the philosophical problem of deciding when we became us. Nevertheless, we would be remiss if we did not permit this field of study to inform our perspective. So we will allow it to have its voice. To this end we draw, largely though not exclusively, from the work of Nicholas Wade (2006).

Darwin was right, humans and apes have a common ancestor. Some people have misinterpreted this to imply that we have descended from monkeys. The record needs to be set straight once again. The lineage goes something like this, keeping in mind that the details are always in need of refinement. Ever the consummate naturalist, Darwin would applaud this tentativeness. At the outset, it needs to be stated that apes differ from monkeys in that the latter come with prehensile tails used for grasping objects; the former do not. We can identify ourselves according to the standard classification scheme:

1. We belong to the kingdom Animalia because we are animals;

2. We belong to the phylum Chordata because we have a tube (a dorsal notochord) in our backs through which the nervous system runs;
3. We belong to the subphylum Vertebrata because the above chord that we possess is specifically a backbone;
4. We belong to the class Mammalia because we give birth to live young who feed at their mother's breast;
5. We are members of the class Primates for we are the highest among mammals, and this division includes monkeys and apes;
6. We are members of the suborder anthropoid apes or pongids;
7. We are members of the superfamily hominoids since we are apelike;
8. We are members of the family hominids for we are humanlike;
9. Our genus is *Homo*, meaning human (the Latin homo and the Greek anthropos, from which is derived the term anthropology, both refer to human);
10. And our species is *sapiens* for we are sapiential, we demonstrate wisdom, although this might be a presumptuous self-assessment.

Once the primate line broke off into the ape and monkey branches, the great apes were set to emerge. In successive branchings subsequent to the above, and in estimated timeframes that continue to encounter reappraisal, the lineage yielded orangutans at about 15 million years ago (mya), gorillas at about 10 mya, and chimpanzees between 8 and 5 mya. The latter marks the last time that humans and other apes shared a common ancestor. Since then another lineage has separated from the chimp line, namely their smaller and less aggressive version known as bonobos, roughly 3 to 1.5 mya. This means that they and we never had direct kinfolk.

Here are some of the highlights of the arrival of humans and this will be recounted in a series of firsts. The first walking apes belonged to the genus prior to our own, known as *Australopithecus*, meaning southern ape. They emerged about 4.4 mya in Africa and their diet was largely plant food, until roughly 2.5 mya when a more carnivorous/omnivorous diet took over. Hence we have almost 2 my of vegetarianism in our history. This first carnivore was *Homo habilis*, though there is debate as to whether it warrants the category of human. On the negative side of the ledger, its body was still apelike and it was not completely weaned from tree life; on the plus side, it had smaller teeth, indicative of a more carnivorous diet, a larger brain as a result (given the nutrients that meat and less energy devoted to chewing, less time eating and more efficient caloric intake afford), and they were the first to wield opposable thumbs. Also on the plus side can be listed their tool-use, which is what *habilis* means, as a type of handyman. They were probably the first to make and use stone tools for purposes of hunting. Whether the plus or minus sides of the issue ought to be accorded greater weight is a matter for the academicians.

Round about 1.7 mya, *Homo habilis* yielded *Homo ergaster*, which might have been the first human to become a naked ape by losing its hair as well as develop dark skin, circa 1.2 mya, in protective response to its exposure to the African sun. (News flash, we all have deep skin pigmentation in our ancestry.) *Homo ergaster* then gave way to *Homo erectus*, which could have been the first to use and control fire for heat, light and food preparation as early as 1.6 mya. They further mark the first exit from Africa, their remains having been located as far east as China. Another group, *Homo heidelbergensis*, produced an off-shoot called Neanderthals. Debates continue to swirl as to how closely related we are to this community, some researchers claiming periodically that Neanderthals and *Homo sapiens* are members of the same species. Allegiances to the yea or nay sides go in and out of fashion. We can even be more specific at this point. There does appear, at last count, to be up to a 4% amount of genetic contribution on the part of Neanderthal DNA to our own, with French, Chinese and Papua New Guinea cultures bearing the greatest proportion. This implies that there has in fact occurred some interbreeding between us and them, which in turn prompts the conclusion that "closely related mammal species may still be able to hybridize, so this was certainly possible between Neanderthals and Cro-Magnons" (early European *Homo sapiens*). And how this Neanderthal input affects us is currently under investigation (Fagan, x-xi; Stringer, 196-97).

Modern *Homo sapiens* have about a 200 thousand year (ky) history. At approximately this time our brain size attained its current volume. By 100 kya the rest of our bodies became anatomically modern. Then by 50 kya we became behaviorally modern. Three significant events occurred at roughly the latter time: language, that is, articulate speech, made its appearance; the ancestral human population formed; and some of these began to venture out of Africa in the second great migration. At about 40 kya arrived another triplet watershed event called the Great Leap Forward. Here emerged the burgeoning of art-forms, such as figurines, jewelry and painting on cave walls; people began to bury their dead with artifacts and rituals, perhaps anticipating an afterlife for the deceased for which they require preparation; and based on the foregoing, a religious consciousness. We were hunter-gatherers for more than 2 my, meaning natural selection has shaped us for this type of lifestyle. Thus evolution has not as yet been brought up to speed and adapted us to agriculturalist and industrialist economies (Ehrlich, 166, 170; Gazzaniga, 310; Morris, 197–200; Wade, 5, 8, 17–19, 23–31).

These highlights leave us with a perplexing two-pronged question, namely when did we become human and what makes us human? We will treat each in its turn. As regards the first line of inquiry, here is a list of the prime suspects. Did we become human when we began to walk upright and adopt sustained bipedalism? If so, then like our australopithecine forebears, humans can have a more apelike anatomy and reside in trees. We are not alone in our vertical posture, for birds and kangaroos can also display bipedality. The same logic applies to the manufacture and use of tools, for *Homo habilis* still appeared apelike and did not entirely forsake tree life. Other creatures, like chimps, bonobos, birds and some insects also use tools, though do not really fashion them. Was it when we harnessed the power of fire? We seem to be alone in this. Or perhaps when we recognized our reflection in water? Well, other apes and even some dolphins and Asian elephants can accomplish a similar feat. How about when we began to communicate through language? Well, bees, dolphins and whales can perform likewise.

Then was it when we thought abstractly, symbolically and represented it through forms of art? Only we appear to have an aesthetic sense, though not always taste, and take pains to decorate our domiciles. (There are some who maintain that certain other primates also enjoy sunsets.) Or when we began to cooperate and work in groups? Yet the social insects—ants, bees, termites and wasps—organize this way as well. Or perhaps when our brains increased in size and complexity and reached their contemporary standard? Volume, however, would not be the deciding issue, for elephants and whales own large brains too. By extension, was it when we became anatomically modern? Or behaviorally modern? When we attended funerals? Perhaps it was when we became agriculturalists, though attine ants also have this in their repertoire. Or when we domesticated animals? Ditto for the ants again, for in a mutualistic/symbiotic way, they have also been known to keep and tend aphids as livestock.

As for the second question, here is another list, some of the items of which do double duty with the first question. The issue as to the inherent differences between us and the rest of the animal world can be subdivided into two, namely differences in degree or kind. The former means that while humans and other animals share certain traits, humans simply exhibit more of the same. The latter refer to those characteristics that do not have a specific analog in the animal world. The difference between these two types can be summarized this way: no matter how many differences of degree one can identify within one trait, no amount of them will ever add up to a difference in kind. Yet if differences in degree occur over a number of traits, the accumulation of variations-become-adaptations will ultimately yield a difference in kind. This is how natural selection operates. A difference in kind is achieved when an organism will no longer be able to interbreed with its parental stock and produce fertile offspring. This signals the onset of a new species.

Differences in degree: As mentioned previously, humans use tools but so do some other creatures. (The difference in kind here is that we also make them. No other species manufactures nuts and bolts, nails and screws, doorknobs and hinges.) We further walk upright, but this is also how flightless birds get around. (We seem to be the lone species with a head propped directly atop our spinal column, though other organisms, like kangaroos, face forward rather than down.) Other creatures possess opposable thumbs, though none with our range of motion. We dance, but so do bees. We make music, but so do the songs of birds and whales. Perhaps we engage in the foregoing two items for their sheer enjoyment, whereas other species do so for communication purposes, sometimes announcing territoriality. Yet the claim can also be made that with songs and rhythmic movements, we also wish to convey attractiveness and availability.

We grow food, though certain ants tend fungus farms. (The difference in kind there would be that only we develop breeds of tulips for pleasure.) We plan for the future, while squirrels store for the winter (yet only we make five- and ten-year plans.) Moreover, we are not the only organism to have a theory of mind, that is, the recognition that other members of the same species have the property of being aware of their counterparts and are able to strategize in opposition to them. They notice when they are being observed, for example, and take steps to avoid detection or otherwise arouse the suspicion of potential competitors. In essence, it is the awareness that the same strategic ability that one possesses is also a faculty in the other members of the group, so one had better be circumspect. And in a related vein, we participate in intragroup/intraspecies organized aggression. We engage in battle with our conspecifics, though warfare is also found in ants and chimps.

Differences in kind: We appear to be unique in using and controlling fire, in cooking our food, in cleaning up after ourselves (though not just in terms of grooming) and tidying our environment (other species might clean themselves but not their place setting or take out the garbage); in the making (though not in the use) of symbol systems; and in symbolic representation in art forms. We practice a skill so as to refine it; we seek meaning and purpose; we are not only aware but self-aware; and we are not only conscious but self-conscious (assuming for the moment that a self exists). Some of these themes will be amplified below. Note that this is a shorter inventory than for differences in degree. What makes us unique, then, is not quite as extensive as we perhaps initially suspected. Despite their lesser amount, however, these are significant differences.

An extended version

Let's take a closer look at some of the distinctives mentioned above in our audit of humanness, in order to assess what it takes to be human. Flexible thumbs that can act in opposable ways may not be common in the animal world, but they are not unknown. Other primates of both ape and monkey varieties display them, but by comparison, ours exhibit gymnastic abilities. Ours can "rotate in great swirling arcs" across our palms so as to connect with the two digits furthest from our thumbs (I was going to say the ring and pinky fingers, but I see no need to resort to technical jargon). This capacity is termed ulnar opposition (all right, so maybe some terms are warranted), and is found nowhere else in creaturehood (animaldom?). And with it we can swing a golf club or put pen to paper. But more than this, we are enabled to exhibit fine-tuned precision movements whereby we can thread a needle. As Chip Walter uncovers, "When picking up something as tiny as a grain of rice, a chimpanzee has to squeeze it between its thumb and the flat of its index finger, like we hold a key or credit card." Our "musculature and nerve structure" allow us feats of manipulation that are foreign to other organisms and are downright acrobatic in relation (Walter, 46).

By the time this talent arose, roughly 2 mya, *Homo habilis* found that it came in 'handy' when forging and wielding tools. The australopithecines from which *habilis* emerged used tools but did not make them. Strictly speaking, we are not even the only ones to make tools, for chimps are seen to make spears to hunt and kill bush babies. Chimps will break off twigs from trees, trim off the leaves and sharpen one end with their teeth. Not high-level technological sophistication, but a rudimentary making of tools nevertheless. Yet still a far cry from an arrowhead. Already decades ago, Jane Goodall observed that chimps also display emotion, though it is difficult to be certain that we are not being anthropomorphic in this. There may or may not be a one-to-one correspondence of apparent similarity to our behavior and the emotions which lie behind it. Furthermore, chimps can imitate the observed behavior of others in order to accomplish a task; they can learn to copy but this tendency does not extend to instructing others to do the same.

Lastly, both chimps and bonobos can cooperate to reach a common goal and thus begin to display an incipient cultural capacity (from the National Geographic program "Ape Genius").

Along with tool-making skills arose others somewhere in our evolutionary development that allowed us to demonstrate the following traits. We are characteristically the lone creature enabled to blush, more noticeable, of course, in our fair-skinned varieties. We are also unique in our capacity to cry. We are not alone in possessing tear ducts, but ours are used for more than ocular hygiene. Tears for us can also be the product of an emotional state. More on this below. We further display moral indignation at the cheating behaviors of others. Some primates voice their disapproval at a perceived lack of reciprocity, but we alone, owing to our language forms of communication, will inform other members of the group as to "who is a violator and who is trustworthy" (Gazzaniga, 31–32). This will no doubt keep informants and private investigators employed.

As mentioned before, other animals sing and produce melodical sounds, but only we create musical compositions, craft musical instruments so as to elicit those sounds, become proficient in their use, master this art and band together into symphony orchestras. No other creature gives such a performance. In relation to us, "Birds sing only in certain contexts: mating and territorial defense. Singing is done primarily by mates, and its sole function is for communication. This also seems to be true of whales. It is not done for pure enjoyment" (Gazzaniga, 233–34).

Some anticipated marks of uniqueness, however, are, perhaps surprisingly, found elsewhere in the animal kingdom. An item of background information, though, before we commence. A requirement for language abilities in humans, for instance, and not found in other primates, is the larynx-tongue-nasal cavity relationship. In chimps, the back of the tongue is still contained within the mouth; in humans, it is exposed to the throat. In the human situation, not only is vocalization permitted but also speech. Yet there is a trade off. "The lower position of the adult human pharynx…means that air and food travel a common pathway behind the tongue;" indicating that we can tell stories, but doing so with a full mouth might present a choking hazard. In the chimp situation, breathing and swallowing can occur at the same time, implying that they are free from the concern of obstructed passageways. Fortunately, human infants reflect the chimp structural arrangement, hence they "can safely suckle and breathe simultaneously" (Ehrlich, 152).

The significance for our purposes is this. Human vocal abilities also "[find] parallels in African Grey parrots." In fact, as Morris argues, "many of the evolutionary features that help to define the human are convergent." That is, such characteristics can arise independently, in cases where the two organisms in question do not share a recent common ancestor. For Morris, such independence suggests that these traits could develop in any general evolutionary scenario. If it can be accomplished on our planet, then no habitable one is exempted (Morris, 226–28). Much of evolution is admittedly divergent, but some forms have found common solutions. As already attested to, bipedality is found also in kangaroos, birds, and in some of our distant extinct predecessors, the dinosaurs (they ruled then, we rule now). Even our ape cousins can at best only muster knuckle-walking, in a sustained way at least. In actuality, most of the features listed above by themselves do not define humans as unique. Even if the traits are more highly developed in us, the difference is still one of degree (Morris, 268–71). It is their cumulative effect that makes us special. Can differences in degree amount to a difference in kind, other than the obvious designation of kind as species? Yes, I suppose, if found in combination. No other creature has this set of higher-order developments. Ours are both more numerous and refined.

Now is the time to focus on our most celebrated difference, our brains. We do not have the largest ones, that distinction is reserved for elephants and whales, so we cannot claim superiority in this regard. But outside of creatures like these, there was a time in our evolutionary history when another organism of comparable size out-distanced us. Up to approximately 1.5 mya, dolphins boasted a greater brain volume. It was not until *Homo erectus* was well underway that we surpassed the dolphins (Morris, 247).

On the topic of present-day comparable sizes, our relative brain volume is "four to five times larger than would be expected for an average mammal." And our brains have even outpaced our own bodies in growth, to the point where our cranium houses about 1340 cc. (cubic centimeters) of gray matter compared to a chimp's 400 cc. In these particulars we are superior, though Darwin recognized that this is still a difference in degree and not kind (Gazzaniga, 11–12).

The fashioning and use of tools also had a marked effect on our brains, and not, as was previously assumed, the reverse. In the process, our brains became reconfigured "so they comprehended the world on the same terms as our tool-making hands interacted with it." Our tactile awareness shapes our neuroanatomy. Even the act of observation has a neural effect, thereby allowing knowledge and techniques to be passed on to subsequent generations through education. When a parent instructs a child on the proper use of a tool, she is able to put observation into practice, for "the same neurons—her mirror neurons—are firing in her head that are firing in her parents." When it comes to be her turn, she can summon "those fired neurons to guide her hands to do something she has never actually done before but *has* imagined doing." And repeated attempts yield proficiency, which can then be bequeathed to another generation (Walter, 58–60). (All well and good, though one difficulty is how technique is transferred to organisms lacking our cranial capacity. Spiders, for example, do not receive schooling in the art of web-construction. Is it always adequate to chime in with "chalk it up to instinct"?)

Such procedural learning, as it is termed, distinguishes us from the rest of the animal kingdom. As Gazzaniga outlines, "To vary or refine a motor movement, one needs to rehearse the action, observe its consequences, remember them, and then" make the necessary adjustments. This is an example of "a rehearsal loop, something we are all familiar with. [But] other animals do not do this. They do not initiate and rehearse actions entirely on their own for the purpose of refining their skill. Your dog is not practicing shaking hands all day while you are at the office" (Gazzaniga, 353).

Big brains mean that we are no longer dependent on genes or at the mercy of natural selection to get our survival work done for us. We are now sapiential and have the resources to think our way through high-level issues (like what to wear to the party, though only some would claim that this is a survival matter). Much of this is thanks to our prefrontal cortex, which is a newcomer to our brains and is placed directly above the bulk of our eyes. This might be why we sometimes massage our foreheads when struggling with a thought or attempting to recall a memory. And the onset of the prefrontal cortex was, in evolutionary terms, quite rapid. The last *Homo erectus* members at about 300 kya did not possess one, but all of us do. Relatively speaking, this development was in the blink of an eye. This "most complex part of our brain," Walter discloses, "is where we do most of our high-end thinking. It is where we worry, symbolize, and process a sense of self and of time, where we recall complex memories and imagine events in the future" (Walter, 103). Whereas bigger brains are a difference in degree, the prefrontal cortex is a difference in kind.

Regardless of the presence of this neuroanatomical structure, some other primates have the ability to identify themselves in a reflective surface, a talent known as mirror self-recognition (MSR). While other species, such as dogs, might view their reflection as another member of their own species, certain primates can correctly understand their reflection as themselves. This capacity is uncovered in their taking advantage of the opportunity to groom parts of themselves, such as their ears, which would otherwise be invisible to them were it not for the mirror. The ability is found in most chimps (though the faculty declines in the elderly), most orangutans, few gorillas and no monkeys. Nor is the skill confined to the primate world, for dolphins and Asian elephants will occasionally exhibit it. Human children also develop the talent around the time of their second birthday. This inventory prompts the question as to why certain individuals within a species display MSR while others do not (a variation without a selective advantage?). The exceptions extend to human "patients with prosopagnosia (inability to recognize faces) [who] cannot recognize themselves in a mirror. They think they are seeing someone else" (Gazzaniga, 310–11; Walter, 114).

Debate also surrounds the issue as to whether a prefrontal cortex is required for the presence of a theory of mind (TOM). Most humans as well as chimps possess the capacity to read thoughts, that is, imagine what someone else might be thinking. We further imagine that the subject of that person's thought is what it might be that we are thinking as well ("Is s/he thinking what I am thinking?"). We can engage in these higher-order deliberations about the mental states of others; a skill which comes in handy when needing to strategize our next move, especially if a surreptitious maneuver is called for. Periodically, we have occasion to "[practice] tactical deception," where we suspect, whether accurately or not, "that in order to trick another individual an animal has to believe that another animal believes something." A TOM begins to appear in humans sometime in their fifth year. At that point they are able to recognize that someone can have a false belief about something, so it would be best to opt for one's own conviction instead. An illustration would be the idea that a person may have marked a box as containing a specific object, when in fact that view is incorrect. By age five, children know enough not to be taken in by the error. Both chimps and autistic children fail to reason accordingly. Some chimps will appear to behave tactically, such as hiding either themselves or objects, based on their perception of the awareness of a potential competitor and the likelihood of their being found out, but there is no consensus as to what extent this skill has penetrated into the ape world. No other organism is seemingly in the running (Gazzaniga, 50–51).

Curiosity is another trait that cannot be touted as uniquely human, for after all, how did the saying originate that its manifestation is fatal in felines? Enjoyment and laughter, on the other hand, has been taken as a distinctly human trait, but can this be ruled out in some chimp and dolphin antics? Admittedly, what laughter accomplishes for us is to solidify the members of our clan, with whom we align and bond. And when a TOM is combined with a clan mentality, some researchers claim that this is where a moral sense surfaces, since it enables us to put ourselves in the place of another (Walter, 131, 144, 161). For then strategy becomes empathy and tactics become ethics. Other creatures may assist members of their group, but some biologists declare that only humans display compassion in the form of "a desire to help relieve suffering" (Gazzaniga, 385).

There are also times when rationality gives way to emotion. In actuality, emotion is indispensable for the decision-making process. Even the choice of which restaurant to dine in is not a rational exercise but is based on emotion, and the outcome will often rest on such sentiments as, "I have a good feeling about this one." Resorting to the contribution of these subjective categories extends even, or particularly, to the moral domain (Gazzaniga, 120–21). Additionally, Walter discovers that we tend to misplace the cause behind a flow of tears. We expect that people shed tears in response to an emotional or stressful situation, yet it could be that crying "is a way for our nervous system to bring us back into equilibrium." Hence, "we don't cry because we are upset,…but because we are trying to get *over* being upset. That may be the real reason why we feel better after we have a good cry" (Walter, 173).

We are also not the only animals to communicate, yet we are alone when it comes to the production of written symbol systems for language. Information is conveyed in the dances of bees, but only we use vocal chords to utter precise sounds which when strung together deliver a detailed message. There are certain, but few, species that can be taught the rudiments of our languages. These include chimps, bonobos and even African Grey parrots. A chimp named Nim could make the association of a sign and its intended object so as to "express simple thoughts," but the skill does not extend beyond the introductory level. That is, "Nim could not form new ideas by linking signs in ways he hadn't been taught; he didn't grasp syntax…They aren't coming up with complex sentences." In order to avoid the conclusion that other creatures are acting merely out of imitation or mimicry, the dolphin holds some surprises. Sentence structure is not beyond their competence. They can comprehend "both syntax (word order) and semantics (meaning), even when the 'words' were presented in a novel order." We spoke already about their sizeable brains and their powers of MSR, implying that "complex intelligence is not the unique preserve of humans." We can now add to the list the manipulation of symbol systems approaching abstract thought. They might not be attending grammar classes any time soon, but elementary language forms are within their reach (Gazzaniga, 57; Morris 252–53, 257–58).

Our brains are further responsible for manufacturing a sense of self, that we and our bodies are continuous. Regardless if there is a self or not, we do come with the sense of one. Damage to our brains can adversely affect this awareness, signifying that this sense is delicate, fragile and subject to injury. As noted, other organisms have also developed a sense of self, which presented itself in humans probably at the time of *Homo erectus*, who "lived life in a twilight zone somewhere between chimps and us, more self-aware than any other creature of its day, but far from deeply thoughtful" (Walter, 113, 118).

One of the items that can be thought deeply about is the nature of this self. Neanderthals potentially devoted some time and effort to this issue, enough to dedicate plots of ground to the deceased. That a self might survive death and continue to exist in some sense might well have occupied a portion of their worldview. The question then becomes whether any other organism has a "notion of essences." The idea that elephants have graveyards is apparently wide of the mark. Evidently, their preoccupation is about ivory, which is to be found "in the bones of their own species," not just their relatives. Hence this example cannot be used to conclude that elephants have an interest in metaphysical concerns or demonstrate a religious dimension. Consequently, they cannot hold membership in a dualist fraternity. This drives one researcher to announce that "if animals cannot form concepts of imperceptible entities or processes, if they do not possess a full TOM, then they cannot be dualists nor entertain the notion of spirits of any sort. These are uniquely human qualities" (Gazzaniga, 268–70).

When all of the votes are in, it is confirmed that we are not the only organism on the planet with intelligence. In addition to those species already mentioned, we can include crows, whales and squid. Several creatures lie along a spectrum regarding the ability to abstract; some are more adept than others, but they can be found on the same axis. The difference in cerebral capacity is one of degree, but what we do with it, especially the cultures we fashion, are differences in kind (Walter, 239 n. 27). No other animal leaves behind a cultural legacy or bequeaths a rich tradition to subsequent generations.

One aspect of human culture has been its religious character, now, along with language, at least 50 ky old. Shamans in hunter-gatherer societies, for instance, likely invoked the spirit world for success in the hunt. After the time of the Great Leap Forward of about 40 kya, religious artifacts were beginning to be created in the shape of fertility goddesses. These as well as other items in the possession of humans at the time needed to be portable so as to accommodate a nomadic lifestyle. Once settled communities began to emerge roughly 15 kya in the Near East and private property became a real possibility, one could aspire to having personal ownership of larger figures, and this led to a form of religion involving an ancestor cult (Ehrlich, 170; Wade, 8–9, 124). Thus religious practice had a 35 ky head start before it became institutionalized. And once oral tradition gave way to written records, the "faith of our fathers" through sacred text could more easily be passed on to posterity.

With all this religion going on, it seems appropriate once again to consult the Judeo-Christian scriptures for insights into our nature, while not at the same time losing sight of the anthropological contributions listed above.

The Bible revisited

The buck stops with God. Should our vast universe ultimately stem from the inspiration of a divinity, then such extravagant creativity could act to bring forth a myriad of forms that exhibit some divine attributes. If God is orderly, then the cosmos would likely reflect it. The law-like regularities that exist make the task of the scientist workable. If God is intelligent, then creatures that display it should eventually, perhaps even inevitably as Morris would have it, arise. Artists put their heart and soul into their creations, which then reveal something of their character. We could expect a similar eventuality on the part of the deity.

The mention of heart and soul provides us with an apt segue, for we are also said to possess them. The Old Testament (OT) description of humans involves the heart, which means the center or the inner individual. Current terminology indicating what the OT authors understood includes "Character, personality, will, [and] mind." Likewise, the New Testament (NT) counterpart regards heart as "the seat of the will…, of the intellect…, and of feeling," in essence, the person. The heart and not "the brain is the centre of consciousness, thought or will," and is a broader category than mind (Banwell, 465).

Soul, on the other hand, is sometimes used in the OT to denote "possessing life" and this applies to animals as well. Soul is often "identified with the blood, as something which is essential to" life and as such is a "life-principle." It refers to differing types of conscious states and also "designates an individual or person." There are noticeable parallels here with heart and spirit, though in the NT soul is similar to mind but distinct from spirit (Cameron, 1135).

Then comes the thorny issue of what is termed the image of God. According to Genesis 1:26–27, humans are said to be endowed with something that, at the very least, allows us to participate in the moral attributes of God. The list of character qualities that describes God includes love, wisdom, righteousness, grace and mercy, and we in our best moments can exemplify them as well. To possess the image means more than this, but it means at least that. The other aspects of God, all grouped under the metaphysical attributes such as independence, eternity and omnipresence, are forever beyond our makeup. To be physical like us means we are barred from these faculties.

We are also able to communicate with the divine. God's messages allegedly get transmitted through revelation and can be received by human recipients. Our messages to God through prayer can also be sent and obtained, with God enabling this two-way exchange. The image is different from the soul in that, according to the accounts, the image is that which allows relationality between God and humans, and souls/minds provide the capacity for humans to think, reason and, with the heart, make decisions of both a mundane and spiritual nature (covering issues from what hand should I use to turn this doorknob, to am I willing to follow God's leading in a specific matter?). As I perceive it, the image permits us to engage in God-sized deliberations, and the soul selects the shape this ability is to take. This is similar to the difference between form and substance, where, at least in the physical domain, one is not found without the other.

Whether we have in mind the image or the soul, for we can neither be pure form nor pure substance (even zero contains meaning), a perplexing question emerges, namely, when did humans first obtain them? Recall from a previous discussion that we do not know when we first became human. The following offers some possible scenarios.

If to be human means to have an image and a soul, then their presence could be traced as far back as the genus prior to ours, the australopithecines. Depending on our definition, if bipedality is the marker of humanity, then we are not the only genus to have boasted it. This would leave numerous species all laying claim to the status of human. The trouble is that all but one became extinct. This exercise could be repeated for all the other trademarks mentioned earlier. If making tools or controlling fire constitutes humanness, then images and souls have a long history. If something else (or more) is involved, then the label of image- or soul-bearer will need to await the onset of the requisite development.

Let's fast forward to about 300 kya when there likely were as many as three human species cohabiting the planet. The last of the *Homo erectus*, the first fire-controlling humans, were beholding the light of day in East Asia. Prevalent too were the Neanderthals in Europe and the Middle East and early/archaic *Homo sapiens* in Africa. After modern *H. sapiens* appeared on the scene, they eventually, much like many (at least in North America) do today, size up Europe and the Middle East as a destination spot, though now for travel and then for relocation. These journeys catapulted them into competition and perhaps even outright conflict with the more aboriginal Neanderthals. (Sound familiar?) The point is that all three were human and could therefore have possessed an image and soul. The question then becomes why did but one survive? There seem to be two options. Either bearing the image-and-soul was not by itself considered sufficiently valuable by God to ensure their survival, or to be human does not by itself mean to have the image-and-soul, but is confined to fewer than three human species.

Was it then at 200 kya, when brains reached their contemporary standard, when we bore the image and soul? And if so, was this feature extended also to the lone other human species around at the time, namely our Neanderthal cousins? Or was 100 kya and our having become anatomically modern the stage when the privileged status was conferred on us? And ditto for the

Neanderthals? Or, finally, when we became behaviorally modern, when both remaining human species likely sported a religious consciousness? The question remains as to how many species enjoyed the benefits, rights, privileges and obligations of image and soul status? If more than one, then why did the other(s) go extinct if imparting an image and soul is such a major investment on the part of the divine? Was it not deemed significant enough to warrant some security coming from God (as I argued in my previous volume)? In this case, to be human does not guarantee divine watchfulness so as to ensure survival. Or if only one species, namely ours alone, then how did we rate? What made us meritorious? Evidently, simply being human was not enough, else other species would have been bearers. What did we do to deserve or earn it?

Alfred Russell Wallace, a contemporary of Charles Darwin and co-discoverer of the theory of natural selection, perceived the human mind as the single entity which the evolutionary process did not produce. Instead, the divinity deposited a mind directly into one of our ancestors at an opportune time. Even if this were to have been the case, the questions are not resolved. Is the soul reserved for *H. sapiens* alone? Wallace was not in a position to consider many of these specific issues. Yet with anthropological advancements, we are. Besides, we could also ask Wallace about some of the logistics involved. Is mentality a heritable trait? Were all members of humanity given the image and soul at a certain juncture? If so, when might this juncture have been? And if only one individual was the recipient, then lucky for us that s/he survived. Or maybe God needed to initiate this on multiple occasions, environmental threats being what they are. Then, of course, we would need to deal with the concern about how the feature was passed on through only one parent. Wallace's assistance in these matters has apparently elapsed.

Moral of the story: bearing an image and soul does not warrant presumption on our part, for it has seemingly been expendable before. Or if we were the only ones ever to have had it, then being human is not enough. In any case, either being human or bearing an image-and-soul does not by itself count for much. Our value and significance might lie elsewhere, perhaps in how we use our human capacities.

In my appraisal, I see the feature of God-consciousness, arriving at the estimated 50 kya time frame, to have been a product of the evolutionary process and an emergent property of humans. One could say that God worked with the substrate and when "the time had fully come" or "the cake was fully baked," God ignited, quickened or fanned into flame what was there already, rather than introducing a foreign entity from without. In so doing, God enabled the organism to cross the line from creature to person. This does not alleviate all the problems associated with the anthropological line of inquiry, nor the accompanying theological concerns, but it does provide a starting point.

The creation accounts in Genesis 1 and 2 are understood as mythological, but that does not mean valuable insights cannot be gleaned from them. Perhaps even God applauds them for their spiritual merit, instead of heaving a sigh that they miss the mark of historical accuracy. We find in Genesis 1, for instance, a divine evaluation of the creation as very good, and this is probably an aesthetic judgment. God apparently stepped back and admired the beauty of God's own handiwork. (Alternative views exist, such as the evaluation being a logistical one, assessing the workableness of creation in practical terms (Foster, 141).)

Then comes Genesis 3. Upon reaching a state of consciousness, it apparently did not take humans long to use this capacity for self-serving purposes. Having witnessed what humans do with their image-bearing faculties, the creation became a little less beautiful. Humans entered the arena of good and evil and had to deal with issues of right and wrong. No other organism has a moral component; they cannot be good or bad. We might say, "bad doggie," but we do not take animals to court (although this in fact occurred in the Middle Ages, owing to the belief that creatures could be culpable). Should our canine pets arouse our anger, they might cower and slink away, not because of guilt but for fear of punishment. They recognize that an owner's (or pack leader's) wrath means potential grief. They may not be able to make the connection between actions and consequences, but the owner's voice and facial expression spell calamity. If the link were to register with them, then perhaps they too are on the way to developing an understanding of the rudiments of cause and effect.

Genesis 3 informs us that the creation is cursed due to our bad moral choices, and Genesis 6 reveals that our inclination is to continue in this status of distorted image-bearers. The accounts imply, contrary to anthropological wisdom, that our death results from this distortion, and that other animal and plant death ensues from the curse, since they are not in a position to be immoral. The biblical picture of a creation groaning under the weight of the curse makes sense only if there was a time, a Golden Age, when there was no travail and subsequent to which events deteriorated. With no anthropological evidence for this, can the metaphor still be helpful? What would the creation be complaining about if circumstances had always been this way (as per Tennyson's diagnosis of nature as "red in tooth and claw")? If nothing has been lost, then there is no curse and no groaning because of it, just the need to grow accustomed (adapt) to new environmental situations, as evolutionary theory would describe it.

Returning to the topic of hearts, the text declares that God knows what is in them. The prophet Jeremiah proclaims that our hearts are not only devious but perverse (Jeremiah 17:9). In my view, the basic human predicament is God-substitutes—those aspects of our lives that constitute our prime focus and driving force that militate against God's authority. These can take the form of anything that occupies our main interest in life: greed, fame and power being three of the favorites. Left unchecked, they can become all-consuming preoccupations. The problem begins when allegiances are shifted away from God to something in the world. There are parallels here with Hindu and Buddhist perspectives, which, recall, refer to the drawbacks of attachment. As the text makes plain, God countenances no rivals. The remedy is a change of heart, a turning back to God, a return from captivity.

The biblical position appears to teach the following. We are made of the dust of the ground, that is, we are constituted from natural materials already in existence and hence we are physical; plus we are fashioned according to a particular mold, namely an image, which was not present in the world beforehand, and thus we are persons. Additionally, the biblical notion of heart, with all its contents such as soul/mind and will, seems to be included here, and consequently we have a moral component. But there is another factor involved. Genesis 2:7 explains that we have been breathed into by the breath of God, giving us a kind of kick start or a booting up of the system. This is sometimes translated as spirit, so a spiritual component has also been imparted to us. The NT book of James, in chapter 2 and verse 26, announces that "the body without the spirit is dead," implying that the spirit is that which gives life or awakens us, a type of driving force.

If this life refers to any and all life, then other animals and even plants must possess it as well, thereby suggesting that the image-and-mind alone makes humans unique in the biosphere. If this life, however, refers exclusively to ours, then what makes the rest of the living world alive? Our type of "living beings" then must be different from theirs. In what sense, though, might this be?

The thread and thrust of Hebrew existence was embodied in the Torah, the Law, for all membership in Israel. The emphasis there is how to live a life of obedience to God, not on what to believe. About the only propositional requirement on the part of the Jewish people was that God exists and that it was God who gave them this very Law itself, all 613 items of it. In fact, the Law commences with a statement on who God is. Yet changing circumstances yield a change in emphasis. As alluded to in Part One, when the Jews were taken captive into Babylon, they encountered a situation in which there were competing myths. When the priesthood recognized that the surrounding cultures each had answers to religious questions, whose appeal could attract their own followers away from the fold, it was pressed into service to become an impromptu theological enclave so as to submit a counterproposal. There was a gap between what the Jewish tradition and the other religions could explain, provoking a response by the priesthood, which felt threatened by the omission.

The surrounding cultures came equipped, through their own literary efforts, with stories and legends as to how they and the world came to be. The Jews then set to work to fill the lacuna. In order to complete the task, they needed to familiarize themselves with the traditions of their captors so as to determine what it was they were up against. This is the policy of "know your opponent." And their opponents were mostly Zoroastrians, a religion that survives through to the present day with about one-quarter million adherents worldwide. The Jews wanted their own heritage to survive as well, so their strategy became one of incorporating from the competing myths what it was judicious to adopt while maintaining a distance with what was non-negotiable. This was for the purpose of making and keeping a distinction between themselves and others. It seemed to work.

Notions of Jewish history likely emerged at this time. Phase one was their own national story, the history of Israel as a nation, found in the book of Exodus. There is debate, however, as to how extensive this was. One side of the debate sees the history as already fully in place, with an oral tradition recounting Israel as having come out of Egypt and experiencing the formation of their own independent nation. The other side suspects that, as Egypt became the villain for the Persian Empire that succeeded the Babylonian and Median empires, it would be expedient for the Jews to establish their history on the basis of Egypt-as-villain too. If so, then their national creation story originates in Babylon, not previously in Jerusalem. Nevertheless, I suppose that a nation can be villainous, or at least portrayed as such, on two separate occasions from the perspective of two different traditions, both the Jewish and Persian.

Next came the need for an account of how nations and humans in general arose. The Jews elected to sculpt their account in terms of God creating humans not out of the divine substance, as in other religions, but from the resources of the material world already in existence. Hence we are the stuff of the world, not the divine. And lastly, there needed to be some rendering of how this world itself came into being. That one God is responsible for all physical things in the world is a direct frontal assault on the gods of Babylon. Their cosmos is populated with them, and they all come under attack from the Jewish priestly pen. For the Jews, there is one God who expects all followers to adhere to a moral code, hence the term ethical monotheism.

Interesting that the Persian Empire did not object to a monotheistic portrayal, as long, it can be suggested, as Egypt is made out to be the bad guy. Persia would insist on the latter, and if this condition is met, then perhaps the rest can be accommodated. Also of note is this: the more that the Jews moved forward in time, the further backward-looking they became. Phases two and three, the creation of humans followed in literary output by the creation of the world, together with phase one on the creation of Israel as a nation, are in reverse chronological order. As the accounts are crafted, they are about periods progressively remote in time. Hence the writing of Genesis 2, on the creation of humans, likely preceded the writing of Genesis 1, on the creation of the world (including humans). This rendition becomes progressively more political, anthropological and cosmological. Maybe as long as phase one was in place, the Persians could tolerate the quirkiness of phases two and three.

Yet these were not the only topics addressed. In reaction to the Zoroastrian stress on a dualistic struggle between good and evil, and the divinities which embody them, a notion of the demonic emerged in Judaism as a nemesis to Yahweh (Israel's God). The difference is that while good and evil do battle for supremacy in Zoroastrianism, the outcome is never in doubt for Judaism, since the devil is always an underling. The concept of Sheol might also have arisen during this period, but some refinements were to occur later. Roughly three and a half centuries after the Babylonian exile, stretching from about 586 to 536 BCE, the Jews encountered aggression on the part of the Greeks, by their domination of the area following the conquests of Alexander the Great, who died in 323 BCE. In direct reaction to the violation of the Temple by the Greeks, the Jews revolted and gained control for a time until the Romans occupied the region after 63 BCE. While the Greeks were in power, they instituted a "policy of enforced Hellenization" during 167–164 BCE, which "outlawed the practice of Judaism." Jews were persecuted, the Temple desecrated, and members of the Maccabean/Hasmonean family "led a revolt." Their triumph in 164 BCE is celebrated in the observance of Hanukkah (Wilson, 267).

Political pressures once again spurred the Jews into theological action; hostilities sparked literary output. There is some indication that there are two halves to the book of Daniel. Chapters one through six are thought to have been produced at the end of the Babylonian exile; chapters seven through twelve are believed to stem from the Maccabean revolt period. The topic of resurrection is broached in the twelfth and final chapter of Daniel. As a reaction to persecution, the Jewish theologians proposed that a God of justice would not leave the righteous to face a martyr's death with finality. There must be, they urged, a time when grievances will be redressed, when God will establish a system of rewards for the righteous and punishments for the wicked. This can occur only "beyond the grave," when there will be a resurrection, not only of the soul but also the body (contrary to dualistic views). The notion of Sheol as "the underworld where hitherto the dead dwelt, in some discomfort," had developed from the time of the exile to the revolt period and became "divided into two compartments": heaven and hell. This evolution of doctrines had the effect of assuaging the hearts and minds of observant Jews in their time of trial. No matter how bleak these events became, so the propaganda went, God would set things right at the close of history. Several of these ideas have Zoroastrian parallels (Watson, 158, 161).

We will return to some of the latter themes at the end of this Part Four. But for now, as a final installment of biblical topics, I wanted to provide a litmus test of sorts about the man Jesus. In my view, there are two camps about Jesus' humanity that commentators can reside in. I propose the following hypothetical situation. If one were to be asked whether Jesus could have understood our language were we to have magically appeared in the Ancient Near East in conversation with him, those of a conservative leaning would probably respond with "Well he's God, isn't he? So he knows everything, including all languages." Those with a liberal inclination would counter with "Well he's human, isn't he? So he would need to learn any language just like the rest of us." We betray our allegiances, our bent, our stripe, through this simple test. (For those keeping track or score, I find myself in the latter camp.)

The nature of science and the science of nature

Most scientists who have reflected philosophically on the issue, as we noticed in Part Two, contend that human nature is yet another topic for scientific study. Even mentality does not escape the rigors of scientific endeavor. We now investigate the extent to which this claim is accurate. But before we do so, some preliminary comments about the scientific enterprise itself are in order.

As mentioned previously, creatures that display a TOM suspect that they can predict how another member of its species will behave "by inferring its psychological state." This anticipation can be applied more widely in the human sphere, for we have expectations about the objects in our environment, not only the organisms in it, and we infer properties on them. Whereas both humans and other animals enjoy commonality in engaging in observing and predicting, it seems that we "alone also try to explain" (Gazzaniga, 252, 262).

This is where science comes in. Science seeks explanations, but no ordinary ones will do. Explanatory statements must be more than merely the stuff of opinion; they must be tested and tried in the arena of public scrutiny. Anybody can say anything they want, freedom of speech and all, but not here in this crucible; they would not be able to get away with it. This means that there are rules of the scientific game that must be adhered to if investigative work is to be considered reputable. There is no written code as such, that all practitioners must sign or swear an oath of allegiance to (outside of physicians), but it does not lack authority for being informal. Peer pressure is now in the form of the peer review.

The currency that science trades in is factual data. One commandment in the informal code is thou shalt not dismiss data if it disconfirms your pet theories. Not all scientists, though, are automatically predisposed to conform to this rule; recall what Genesis says about the inclination of our hearts. We need to be on guard and check ourselves for any agenda-driven elements. We might well ask whether it is even likely to entertain a non-politically or non-ideologically-motivated science on issues such as global climate change, racial differences or homosexuality as having a genetic basis. The history of science demonstrates that science gets it wrong periodically, but eventually it is self-correcting. The undertaking of science is not intended as a validating experience; it is not for the squeamish (or the dishonest). This calls on practitioners to be dispassionate. Only one passion is allowed: "disinterested curiosity—the passion to know the truth" (Ducasse, 146–47).

There is concern, however, about this notion of truth, and we can come at it from the following angle. Science examines physical causes, so, naturally, there must be a natural explanation for all that it studies, for it does not study anything else. The explanatory power on the part of science has led some scientists in unguarded moments, somewhat hubristically and hastily, to declare that there is nothing other than the physical to study anyway. This is known as the philosophical, note not scientific, position of materialism. In so doing, these scientists have strayed from the safety and security of their scientific confines to wander into the philosophical forum and engage in speculation. There is no problem with this, unless they call it science. A labcoat will not help them here. We can all wax philosophical (with varying degrees of acumen). Even if the entire scientific community were to reach a consensus on a philosophical matter, that will not settle the issue or make it science.

Some thinkers, like Curt John Ducasse, had harsh words for this hubris. The very idea that physical events *can only* be caused by events themselves physical "is not a known fact but merely a metaphysical dogma." Some scientists, not all, claim that a nonmaterial cause for a material event, or *vice versa*, constitutes an impossibility. Ducasse is bewildered as to how a scientist could judge that something is impossible. What he uncovers is, rather, presumption and sloth on the part of a scientific subgroup, who really mean to say that a non-confirmatory phenomenon would undermine all the work they have put in and they would need to rethink, retool and reformulate much of what has been previously achieved. It's just easier to say that it can't happen. (The scientific version of the religious "comfortable pew," I suppose, is the "comfortable review.") The assumptions and assertions on their part, on the contrary, have "not in fact been established by science," and Ducasse is rightly contemptuous of them as sheer dogmatism (Ducasse, 84, 148–49). Dogmatism, it appears, is not restricted to the religious domain and is unwarranted no matter where it surfaces.

Besides, science is not built on conjecture. There are only so many "perhapses" that science can tolerate before it demands the evidence that can decide the issue. The trouble is, this evidence might not be forthcoming. If science cannot explain something at present, the appeal is sometimes made to what can be called a "promissory note," that is, difficulties "are simply deferred to future science." The belief here is such that science is omnicompetent and can thus address all concerns in the course of time. This attitude amounts to an article of faith, a hope in the prospects of an unseen future. Why then is there objection when religionists do the same? Some commentators are even more direct: "We regard promissory materialism as superstition without rational foundation…[It] is simply a religious belief held by dogmatic materialists…who often confuse their religion with their science" (Beauregard & O'Leary, 28, 47, 125 (quoting Sir John Eccles & Daniel N. Robinson)).

Back, finally, to the subject of truth. The thought of cognitive scientist Steven Pinker is singled out for criticism. He rhapsodizes, for example, that "Our brains were shaped for fitness, not for truth. Sometimes the truth is adaptive, but sometimes it is not." Precisely upon recognizing this, some critics are quick to point out, why then do those researchers such as Pinker (and there are several others) fail to apply the observation to their own thinking and, instead, falsely assume "that their ideas have escaped the necessity of evolution and therefore have independent validity—but that their nonmaterial opponents' ideas have not?" (Beauregard & O'Leary, 122) Applying the apt words of Albert Einstein to this end, "there is no privileged frame of reference in the universe," not even for Pinker and his companions.

Having touched on some of the limitations of science, it must be admitted that it contains much in its arsenal. The call for humility on the part of science does not imply that there is not much about which it can justifiably be proud. Even if the divine were to have created the world and is sustaining its ongoing processes, science has unlocked this divine gift of nature. The world is its oyster.

The scientific revolution began in the seventeenth century and the Enlightenment in the eighteenth, but the spirit of rational and empirical inquiry has been around far longer. Biblical criticism, the area of academic study that examines the Bible as it would any ancient literary and historical document, had a significant effect on religious beliefs, and as a method of biblical scholarship it has, like the scientific revolution, an almost four-century presence. Yet its history can be traced back even further. "The first major attack on the scriptures had come as early as the twelfth century, when the Jewish scholar Aben Ezra challenged the tradition that Moses was the author of the Pentateuch" (the first five books of the OT) (Watson, 523). Hence Hebrew canonical scholarship predates the Christian West's counterpart by nearly half a millennium. (What can we expect once the Muslim world's Qur'an criticism takes root?)

In Part Two, we noted how historically there were scarcely two major philosophers who deemed that human nature is not a subject for scientific scrutiny, namely Descartes and Kant. This is because, for them, mentality does not behave like the objects that science investigates, for the mind is not mechanical. For Kant, "There cannot be a 'science' of psychology because what we observe in our minds does not exist as objects knowable in terms of…space and time" (Watson, 534 (quoting Alfred Cobban)). This further suggests the inconvenient implication for science that some knowledge can be derived through non-scientific means.

The bulk of later philosophers did not follow Descartes or Kant in this. Hume, especially, became the champion of the empirical approach, though not in the way that science would hope. Hume undertook religion as an historical study, and this disclosed for him that religion is not sufficiently distinct from any other human activity and as such ought not to be accorded special status. His evaluation rendered religion nonsensical. But the nonsense does not end there, for "Hume thought reason was completely in thrall to passion, and to that extent all science was suspect. There are no laws of nature, he said, there is no self, there is no purpose to existence, only chaos." What is worse, in our attempt to order life experiences into a meaningful whole, a natural human tendency, our "knowledge becomes belief, 'something felt by the mind,' [and is] *not* the result of a rational process" (Watson 539).

When we tabulate the philosophical results, we find that we cannot adequately have confidence in either science or rationality, since they are both contaminated by human emotive aspects. As long as scientists are human, and this was the error of the philosophical movement known as logical positivism or the Vienna Circle, there will always be human subjective elements in science. Thus science cannot always expect Hume to be an automatic ally.

Life as a human

We have completed our look at human origins and early history, scriptural contributions and the ground rules of what science can and cannot accomplish. The time has now arrived for us to concentrate on the biological sciences for more clues, since the study of human nature will, at least in part, benefit from and be illumined by scientific investigation, despite Cartesian and Kantian objections. Having examined the development of humans as a species, we now focus on the development of humans as individuals and consider what makes us human and what humans are like "from the womb to the tomb" and even beyond.

We begin life as a single-celled organism in an intrauterine environment. There is not long to wait until the single cell becomes several, the one becoming the many, which in turn yield a one once again, a single individual (or more should the egg(s) comply). We become noticeably human in form as cell division proceeds and gene regulation allocates cells with specific functions to specific addresses. This is why some cells, all with the same genetic complement, become hand cells and others foot cells.

As concerns the life of the human fetus, the intrauterine environment may have no effect on the neonate to be delivered, but research suggests otherwise. It appears that "the behavior and experiences of the fetus" actually does have an "impact on its development" and is significant "both before and after birth." In particular, the brain encounters its greatest rate of growth during the fetal stage. "The prenatal period marks" the highest production of neurons in our lifetime, with about one-quarter million brain cells manufactured each minute "[a]t its peak," prompting some researchers to urge that this phase of our lives "may provide necessary and essential stimulation for the formation of the CNS [central nervous system] and subsequently its function." Even the psychological state of the mother can adversely affect the womb, the consequences of which extend through to adulthood. In terms of our development as individual persons, the importance of our life prior to birth should not be discounted (Hepper, 474–77). Formerly, this period was understood as insignificant, though some Asian cultures have long recognized its meaningfulness by regarding newborns as already having reached their first year of age upon delivery.

In the case of twins, genetic sameness does not necessarily equal sameness of natures, despite their fetal environments as having been much the same. Even though they were in close proximity in the womb and even if they were to have been treated the same in their formative years by primary caregivers, "Quite subtle environmental differences, perhaps initiated by different positions in the womb, can sometimes produce substantially different behavioral outcomes in twins" (Ehrlich, 10). Minor differences, especially their cumulative effect, can combine into major ones.

Size or complexity of an organism is also no indicator as to the size of its genome. As a corollary, the amount of DNA a creature houses is not directly proportional to its complexity, for even "single-celled organisms may contain far more DNA than a human." The bulk of the genetic material might fail to accomplish what DNA ordinarily does, other than the aforementioned regulation, namely code for protein. The large amount of surplus DNA, in excess of that which is active in coding, is referred to as "junk DNA" (Morris, 237), its purpose and function as yet still undisclosed. The roughly thirty thousand active human genes, a paltry amount relative to some far less complex creatures, comprise "a little more than 1.5 percent of the whole genome" (Gazzaniga, 14, 33, 42). Apart from our ecological policies and frames of mind, we begin life, as all organisms do, being intrinsically wasteful.

Now that we have entered the world as individuals, though far from independent—we would not survive without frequent attention—the study of the human is about input and output. With regard to the latter, genes have a definite influence on our actions: "genes clearly constrain human behavior and possibilities, providing dispositional tendencies for the organism; but they do not entail or determine all the behaviors of the organism over its lifespan" (Clayton, 98). There are also what are known as automatic processes, which come in two varieties. The first has output features and is called intentional. Here behaviors are inculcated and become second nature. Education assists in the learning of these goal-directed processes. The second is input-oriented and is termed preconscious. We are built in such a way that stimuli are perceived in a preprocessed fashion: "your brain processes [a stimulus] before your conscious mind is aware that you have perceived it. This takes place effortlessly and without intention or awareness" (Gazzaniga, 121).

The next step takes us to the borderland of science. When the topic of discussion is whether human nature comes under the rubric of science, the following stage is patently peripheral to it. We have an interiority or subjectivity which does not play by the rules of science or submit to its authority. It conforms to a different regimen, for we "have nonperceptual psychological properties not subject to physical laws." Subjectivity is a world that science is ill-equipped to address. Possessing interiority, "We understand that there are invisible forces. Current evidence suggests that we are the only animals that reason about unobservable forces. We alone form concepts about imperceptible things and try to *explain* an effect as having been caused by something." Curiosity, the impetus for discovery, is also found in other creatures, but the drive to organize findings to fit into some type of order or scheme is both unique to humans and irreducible to scientific formalization (Gazzaniga, 274, 384). Our tendency to collect and categorize likely stems from our long history as hunter-gatherers.

Subjectivity also finds its way into the thinking process, for there is a distinction to be made between the electrical activity in the brain and the thought that it generates. They are eminently not the same thing. By the above reasoning, the former is scientific, whereas the latter is not. Neuronal activity can be observed in perpetuity, but this will not enable us to capture a thought. Thoughts and feelings themselves are obviously unobservable. None of us is without subjectivity, in the sense that it is objective (pervasive) and undeniable, but it is not scientific. Nor can we claim that mentality is a function of, or even resident in, the brain. Yet, contrary to all non-dualistic positions, each of physicality and mentality affects the other. In the physical to mental direction, indigestion adversely colors our outlook on life; and in the mental to physical direction, "fearful thoughts increase the secretion of adrenaline, but happy thoughts increase the secretion of endorphins." There is thus a two-way (unknown) mechanism of information translation (Beauregard & O'Leary, 150–51).

So a funny thing happened on the way to full cranial humanity: mental events can causally influence neural structure. Here is the extent of the effect: a cardinal belief in the early days of neuroscience was that adult brains undergo no change. The discipline now recognizes that neurons continue to "reorganize... throughout life, not only in early childhood." While no new neurons are created in the adult stage, "Our brains rewire to create new connections," a phenomenon termed neuroplasticity, where "synapses constantly form and dissolve, weaken and strengthen in response to new experiences." This means our decisions (mental) affect our neural circuitry (physical). If we choose to curtail a certain behavior, such as smoking for instance, new dendritic connections will form to accommodate the change, and with repetition will become reinforced, or, in Rupert Sheldrake's view, carve an even deeper groove in our morphogenetic field to the point where not smoking will become habitual, or our default drive. Since our brains are plastic and respond to the traffic we give them, "if we change our mind, we change our brain." This "has been demonstrated by experiments and is even used in psychiatric treatments for obsessive compulsive disorder" (OCD) (Beauregard & O'Leary, 33, 102–03).

Genetic predetermination, therefore, cannot account for the ongoing changes in our neuronal network. Instead, our brain "grows epigenetically.... It is formed by our actual encounter with reality." This occurs to the point where the circuitry will be rerouted in response, say, to an amputation. "If an amputee's brain has not changed its mental map of the body after the amputation, she will experience those feelings as if they came from her" phantom limb. Rather than remaining stuck in the initial format, the neurons will rewire to form connections of dendrites that will deal with the new situation. New pathways can thus be forged. "[N]eurons that once received input from a vanished hand could [for example] rewire themselves to report input from the face." Additionally, neurons carrying stimuli from formerly functioning eyes might regroup so as to enhance the sense of hearing, or "Part of the cerebral cortex devoted in sighted people to vision are recruited in blind people for tactile processing," thereby facilitating the mastery of Braille (Polkinghorne, 48; Beauregard & O'Leary, 103; Foster, 249 n.21).

I have a comment to make at this juncture and it is a theological one. It could be stated that there is a neuroanatomical record of the "deeds done in the body" through actions having become habits, and these are located in the dendritic pathways formed as a result of our decisions. Choices that are repeated often enough become habit-forming, which are then more reflex in nature than conscious decisions. They have become firmly etched or entrenched in our gray matter. If we ever elect to change our ways, as God has been known, at least biblically, to prompt, then what we must face includes the physical structure of the brain that we ourselves have established through our choices. Decisions form new pathways, but habit, as the saying goes, is a formidable force and tough to break.

Now that we have some indication as to how the brain adapts to changing circumstances, more is needed about how it works and how we use it. Most significantly, we are bombarded with information, only a minute portion of which we can sift, weed, screen and filter. To attempt to digest the constant barrage of sensations "we encounter…would leave us paralyzed." Thus we must constrict the sluice gates in order to ensure that only the most salient features of our world of data register. Chris Hedges makes the conundrum plain: "We are assaulted with about 14 million bits of information per second. The bandwidth of consciousness is around 18 bits per second. We have conscious access to about a millionth of the information we use to function in life. Much of the information we receive and our subsequent responses do not take place on the level of consciousness" (Hedges, 160–61).

So this is what our brains are up against as well as the adjustments they make. Hedges has more to say about complex brain functions and it seems to be in line with Alasdair MacIntyre's notion in Part Three concerning self-narration. We make reference to ourselves in narrative form; we inform others about ourselves through the use of stories. Narratives allow us to make sense of our lives, of combining disparate thoughts and events into a unity, without which there would be only "incoherence and fragmentation." The trouble is that "Our self is elusive. It is not fixed." Nor has its presence ever been established. Nevertheless, we concoct stories in an effort "to give structure and meaning to our life" (Hedges, 159–60). Hence these tales are inventions; they are manufactured. And we can be even more specific about which part of the brain is responsible for them. "The left brain constructs a story to help explain the actual behaviors that are pouring out of the right brain." Yet as stories, they do not necessarily reflect reality. At all times they are a fabrication (Walter, 135–36 (quoting Gazzaniga)).

Personal narratives also have a usefulness outside of individual apologia, namely their application makes them "the source of our earliest myths" for a community. For purposes of a corporate body, they provide a common story that the group can rally itself around and give itself an identity. And when fully elaborated and articulated, myths take the form of a metanarrative, fit for a longstanding cultural tradition (Walter, 138).

A comment now on the relation between intelligence and emotion. The assumption is made that the more one has of the former, the less we will need or retain of the latter. In essence, rationality is there to mop up any residual emotive spillage. Walter is emphatic about his rejection of this view: "Our emotional life is more complicated and enriched *because* of our intelligence, not because our intelligence has obliterated our less intellectual side. In fact, our big brains have *created* the immense emotional life we all enjoy." The actual state of affairs is such that "our intellect amplifies the primeval parts of us. A storm in our brain is not a bad analogy" (Walter, 139, 199). We will not eliminate our emotional side through intellection; nor is it obvious that this would even be a legitimate or desirable pursuit.

A final note on natural selection itself before we embark on our next topic. Not all traits fall into the category of adaptive or maladaptive, for some can be adaptively neutral. These are the kinds of features that natural selection has not as yet selected for or against, since they have not become issues in survival or relevant for reproductive success. An example is the onset of age-related ailments. Only those traits which surface during reproductive years enter into the adaptational sieve, since only then are they factors in fitness. Characteristics which appear most in the elderly, such as osteoporosis or spinal cord compression, do not figure into whether an organism leaves offspring, for the reproductive stage has likely already passed. It has not been long in human history that we have been able to make this observation, for we now encounter a longer lifespan than ever before. For the senior phase of our lives, natural selection doesn't have much investment; in that period we are biologically irrelevant, except for the care of grandchildren so that they too might become educated and reach reproductive age.

Awakenings

What we can glean, perhaps even what provisional conclusion we can draw, from the previous section is that the left hemisphere of our brain is responsible for at least two human tendencies. The first is to impose order on our environment, but not only externally, for we seek coherence internally as well. Our picture of the world and not just the world itself must be subdued. We organize thoughts in addition to living spaces; we look for patterns and construct hypotheses in order to unify what we see and know. We are dissatisfied with data alone; we must integrate them into a framework that makes sense of our interior and exterior worlds. We are theory-builders. The second is the creative task of forging a running commentary of ourselves and our place in the world that we construct mental images about. We pursue unification here too, the product of which is a sense of self who has a story to tell about itself. This part of the brain might recognize the functions it performs and posit itself as the being or agent that performs them. In essence, the one who interprets these images is the one we refer to as me. Thus what we call "the self is a knowledge structure, not a mystical entity." Only humans ask about meaning and purpose; only we give ourselves a narrative; and only we have the point of view of being observers and not just participants (Gazzaniga, 294–97, 300–03). And with the left brain, the construct of the self is awakened.

The comparison was made in Part Three between computers and the human brain. The shortcoming of computers is that they "don't form...plans, nor do they have goals. They do not have overarching ideas, nor do they use analogy or metaphor—and there is no way currently proposed to make them do so." This perspective might be a touch hasty, however, for the computer named Watson, when on the game show Jeopardy, was made to form connections that verge on cleverness. Perhaps this holds promise for future progress so that the AI frontier can be tamed after all. Computers are not smarter than humans, they are faster. Despite this fact, here is another: the human brain has far more connections than any computer (Beauregard & O'Leary, 22; Gazzaniga, 358, 361). So what is to be deemed superior, speed or expansive thinking?

On the topic of these cerebral connections, prior to the onset of brain cell deterioration, which occurs throughout our lifetime but is most pronounced in the elderly, there are about one hundred billion neurons in the human brain and each of them connects to roughly one thousand others. If we do the math, this gives us approximately one hundred trillion synaptic connections in our cerebral cortex. The number of possible pathways which signals could take in this network not only outnumbers the stars in the universe but also the atoms (Barbour, 142). This gives us an idea as to the complexity of our brain. But even such a structure can have its glitches. For example, "if certain stimuli trick your visual system into constructing an illusion, consciously knowing that you have been tricked does not make the illusion disappear. That part of the visual system is not accessible to conscious control" (Gazzaniga, 283). Such counter-instances can be multiplied. Hence the brain is powerful, but it is also delicate. With much sophistication comes much that can go awry.

Consider another example. When dealing with the phenomenon of neuroplasticity, we mentioned the condition known as OCD, the main concern of which "is that the more often the patient actually engages in a compulsive behavior, the more neurons are drawn into it, and the stronger the signals for the behavior become. Thus, although the signals appear to promise 'Do it one more time and then you will have some peace,' that promise is false by its very nature. What was once a neural footpath slowly grows into a twelve-lane highway whose deafening traffic takes over the neural neighborhood. The challenge is to return it to the status of a footpath in the brain again" (Beauregard & O'Leary, 128–29).

Neuroplasticity facilitates this reduction. Those with religious sensibilities would find parallels here with what could be called "spiritual warfare," the battle of taking one's cues from God versus, as the Bible refers to them, "principalities and powers." Interesting that the effectiveness of this strategy has a biological counterpart, for, as was noted, changing one's mind changes one's brain. Turning to God then means changing one's habits. Not only is there two-way traffic between physicality and mentality, but also, seemingly, between religion and science. The two might not be so far apart after all.

We will return to the body/brain-mind problem again in a moment, but now that we have broached the topic of religion, the following statement needs to be made in relation to the brain: "There is not just one part of the brain that is used in religious thought; there are many areas that come into play. People who are religious do not have a brain structure that atheists and agnostics do not have." Nor can the brain be relied upon to steer us in the proper metaphysical direction, for it has been described "as a machine for winning arguments, not as a truth finder" (Gazzaniga, 143, 151).

Speaking of metaphysics, we have the presence of mind to ask if there is a mind present. This is where the mind-body/brain theme can be reintroduced. Steven Pinker, an otherwise materialist thinker whom we have met before, admits that one of the main difficulties or Big Questions of cognitive science is explaining how neural activity generates subjectivity. Electrical activity lies behind thoughts and feelings but is not equivalent to them. Pinker observes that "Electrical stimulation of the brain during surgery can cause a person to have hallucinations that are indistinguishable from reality." Now, of course, we need not go to these lengths to hallucinate, for they can also be produced through fasting or narcotic means. Nevertheless, the point is that physical measures can be administered so as to alter mental experiences. What we have just described is an example of cause and effect in the physical to mental direction, but the reverse can occur as well. If we are prompted to retrieve a particularly painful or joyous memory, the same brain areas are activated as if we were reliving the actual experience. There is no neuroanatomical difference between living and reliving. Recollections, therefore, can give us inroads and insights into our emotional side (Pinker, 3; Beauregard & O'Leary, 151).

As we observed in Part Three, several positions have been proposed to address the interaction, if such there be, of the body/brain and mind. Let's focus on one as a representative. Epiphenomenalists, recall, would hold that the fictional halo of a saint is a production of his or her saintliness, but it does not itself produce anything; a halo is always an effect and never a cause. Thus the body is only ever a cause and the mind only ever an effect. For epiphenomenalists, the relation of mind to brain is analogous to sparks from an engine or foam on the water, neither of which react with that which produces them. The analogy breaks down, though, since both sparks and foam are physical whereas mentality is not. Indeed, "the spark and the foam are *fragments* of the machine and of the wave, but states of consciousness are not fragments of cerebral tissue." The trouble with attempts to reconcile alleged body/brain and mind interaction is that proponents believe their own approaches to be comprehensive. In "adopt[ing] the scientific program," epiphenomenalists understand their viewpoint as unassailable. Theirs, however, is not the lone position to draw on science. For those in the materialist camp who claim that no such interaction occurs in either direction, for frankly minds do not exist, appeal is sometimes made, as we saw on a previous occasion, to the scientific law of conservation of energy. Though even this does not provide the hoped-for bulwark against criticism, for "is not known,…, to be true without exception," and as such this law would better be described as a postulate since the notion that the universe is entirely a closed or isolated system is itself an unverified and possibly unverifiable assumption (Ducasse, 75–77, 80, 107). Compounding the problem is that part of the universe known as the quantum world does not always operate as neatly as we would like. (For those comfortable with technical terminology, the energy required for an electron to change orbitals does not have the ordinary Newtonian energy effect for, contrary to Albert Einstein's wishes, rulers and clocks do not help us when electrons take no time to make the trip and are never at any point in between the shells. Hence in this realm, $F = ma$ does not hold.)

Reintroducing another previous theme, one phenomenon found in medical science that is detrimental to the materialist posture is known as the placebo effect. Here a patient believes that s/he has powerful medication administered to him or her that will combat the malady that s/he is encountering. In truth, the drug is not effectual at all but might be as innocuous as a sugar pill. Nevertheless, the patient will often be seen to recover from the affliction. So what was it that proved to be the causative agent in the recovery if it was not the "medication"? The lone surviving factor, outside of prayer, itself another subject of study, is the aforementioned belief. All the patient has and can bring into the recuperative calculations is hope, a conscious mental state and nothing more. In effect, the will to recover. But these states, by their very nature, are not testable or quantifiable and are thus not scientific. Consequently, something other than the physical is curative. If the ill person "is convinced that [the pill] is a potent remedy," studies have shown that this can be a sufficient condition for healing to occur. The expectation demonstrates that the efficacy of "inner resources is real" (Beauregard & O'Leary, 126, 141).

There is also a reverse side to this effect; that is, it can work in the opposite way. Whereas the term "Placebo means 'I will please,'…nocebo means 'I will harm'." Patients who fear that harmful substances have been applied to their course of treatment will tend to show negative health effects. This seems to indicate that some responsibility for our wellness or lack of it rests with us. Perhaps Jesus was on to something in his healing ministry when the gospel accounts have him declare, "your faith has made you well" (Matthew 9:22; Mark 10:52; Luke 17:14, 18:42).

There is a concern here on the part of the pharmaceutical companies, not that patients are recovering, but that there is no money to be made on placebos, for "Hope cannot be trademarked," and the corporations are not in the business of marketing sugar pills. Besides, the notion that drugs alone are clinically effectual appears to be undermined (Beauregard & O'Leary, 145, 149).

Beyond awake

Now that we have touched on humans in their conscious state, we turn to states other than the usual ones. We begin with what occurs when we are asleep; two examples of my own will follow in the Appendix.

Dreams. Humans have them. And judging by the movements which some dogs exhibit while sleeping, they are familiar with some type of them as well. Being asleep is not an unconscious state but rather an altered kind of conscious one. We are able to make this assessment because reasoning does not depart from us when we dream. There is a form of coherence to our dream plots, though of a different sort from when we are awake. There are some things in our dreams that we might take as perfectly logical, but turn out to be nonsensical when reflecting on them later. A case in point on a personal note is face to face physical encounters and conversations with my now deceased parents. Makes sense in my dream world, but impossible in my waking world.

Some dreams are so real and life-like that, while in them, we do not distinguish between the real and the imagined. We assume that we are awake in our dreams and are in complete possession of our faculties. Our visual, auditory and tactile senses appear to be operative, although perhaps less so for taste and olfactory. We make decisions in our dream state, though maybe not always the same ones we would be inclined to make when awake, and these are conscious activities by any reckoning. Hence we conduct business as usual in dreams, despite the content potentially varying from the waking state. The situation is markedly different in lucid dreams, however, where the dreamer is fully aware that s/he is asleep and wittingly participates in the dream scenario. This engagement might involve playfulness with the dream parameters and one may even perform a battery of tests within it, like "let's try this and see what happens." Curiosity, evidently, is also a variable in dreams. Yet the rules of the game seem to be different if we are imagining or reflecting while awake. Whether we play a more active or passive role and take more liberties in our dreams versus waking reflections, I suppose, depends on the personal predilections on the part of the experiencer/author (Broad, 156, 162–63).

Perception of physical objects in our data world of experience produces stimuli in our senses. Perception functions differently, however, in our dream world, for there are neither physical objects to perceive nor senses to stimulate. Objects in our dreams are not physical, nor do they arise from physical things. That is, actual physical tables are not required to conjure up corresponding table-like images in dreams as in a cause and effect relationship; we can imagine our own version of tables thank you very much. Yet this does not imply that they are, or at least seem, unreal, for while in the dream state, our brains fail to make the distinction (Ducasse, 128–29).

Dreams are not the only occurrence that make us markedly different from other species, and this brings us to an aside in our discussion. Humans are not only dreamers but brooders. We aspire to and strive for a future plan for our lives, yet we can also despair about our eventual demise. We know that we have a limited time span; our living will eventually give way to dying. Only humans can be nihilists and existentialists. Perhaps there is a natural selection hand or design in this, for the impetus to reproduce can then be driven by both instinct as well as a sense of personal urgency, that is, both nature acting materially and nurture acting psychologically, all in an effort to impress upon us the value of leaving the world with the next generation. Or are we giving natural selection too much credit, here or elsewhere? Are we asking it to explain more than it is capable of? As we mentioned already, we delight in and are enriched by music, but is it appropriate to claim that it thereby has utility as survival value? As a further digression, genetic similarity to other creatures is not analogous to our similarity with each other. We share more than 99% of our DNA with chimps and 99.9% of it with Beethoven, "but that fact [alone] carries no implications of a close correspondence between our musical abilities." Back to our issue. Nor does it seem that we require any mathematical skills as survival tools beyond a bit of arithmetic, a pinch of algebra and a dash of geometry. Yet we also engage in quantum mechanics. While we may need "to make certain kinds of simple logical association," such additional "rational feats go far beyond anything susceptible to Darwinian explanation." Unless, of course, our survival is enhanced because of them, and sometimes it is (such as knowing why radioactivity and ultraviolet rays are harmful and then taking steps to avoid them). But the beautiful designs generated by the Mandelbrot set do not serve to augment our survival (Polkinghorne, 38–39, 45).

We now turn to the topic of religion while not losing sight of the brain. I am not hereby making a thinly disguised appeal for faith and reason, but rather the guideline that in this discussion, religion will largely be in reference to neural activity. The behavior of religious observance has been shown to occur across cultural history, but this does not imply that its inheritance is genetically based. "What humans actually inherit is the *capacity* for abstract ideas like God, the future, ethics, free will, death, mathematics, and so forth." Nevertheless, Karen Armstrong declares that the disposition "to cultivate a sense of the transcendent may be *the* defining human characteristic." Hence one might say that the propensity for religiosity inheres in humans; that it will automatically be expressed, though, is not a given. In societies which boast peaceful relations with other nations as well as a system of equitable welfare, as in Scandinavian countries, or where there is a lack of internal aggression, as in Iceland (populated by descendants of the Vikings who are no longer inclined to express their bellicose side and could be called, I suppose, the bonobos of the human world), there is less motivation to pursue religious matters. (Engagement in spirituality should not depend on economic status, yet sometimes there is an indirect proportionality. The OT book of Proverbs wisely notes that those with plenty have been known to deny God (Proverbs 30:8–9). That's gratitude for you!) Since there is a genetic component to religious activity, in that it is "embedded in [our] neural circuitry," it will continue to surface (Armstrong, 19; Beauregard & O'Leary, 45; Wade, 270).

Dean Hamer wrote an influential work entitled *The God gene*, developing the theme of religious behavior as hardwired in our brains and attempting to isolate the specific cerebral region responsible for it. His effort has been criticized in the interim because "although the temporal lobes appear to be implicated in the perception of contacting a spiritual reality,…, they are not a 'God spot' or 'God module'." Mystical experience as a religious subcategory, for instance, is much more widely distributed than this. Several brain regions are involved in the alleged mediation between the human and the divine. Neurological experiments have been performed on nuns in their meditative states, despite the objection that they merely suffer from the affliction known as fertile imaginations. This claim can be countered: "a person who is 'faking it' should generate a lot of beta waves (typical of strenuous conscious activity) and not many theta waves (typical of deep meditative states)." The nuns were guilty of the latter. "It turns out that there are some things you just *can't* fake." Mystical contemplation, with its purported encounter with the transcendent, thus occurs separately from the white-knuckle activity of thinking. Moreover, the "slow brain-wave pattern [indicative of theta waves] is not unique to the Christian tradition; it has been found in Hindu yogis and Buddhist monks, so it appears to be characteristic of mysticism generally" (Beauregard & O'Leary, 76, 265, 272, 339 n.36).

The point to be made is summarized by researcher Mario Beauregard: "when the nuns were recalling autobiographical memories, the brain activity was different from that of the mystical state. So we know for certain that the mystical state is something other than an emotional state. The abundance of theta activity during the mystical condition clearly demonstrated a marked alteration of consciousness in the nuns." The discussion now leads us in two directions: memory and mysticism. As for the former, Ducasse makes the distinction that memory "is a capacity, not an occurrence; whereas *a memory* is an occurrence, not a capacity." Armed with this distinction, we must caution the intrepid investigator that memory is limited and can be as unreliable as any of our senses. As go our powers of observation, so too goes our subsequent retrieval of them through memory. Both are colored and distorted through our interpretive framework, given to us by our belief systems. We see what we have been trained to see and have grown accustomed to seeing with the help of our educational and social upbringing, and we recall in a selective way what best fits our agenda. We believe that we see clearly, but "our nonreflective intuitive beliefs" can be faulty. Memory content is only as good as sensory input, and at times far less reliable. Courts of law and the ability of witnesses to bear testimony run up against this very problem (Ducasse, 305; Gazzaniga, 273).

As for the latter, and as already intimated, there may be a gap, critics have spotted, between authentic and illusory experiences on the part of the mystic. Here too we run the risk of faulty interpretation. As Beauregard maintains to the contrary, "There is no scientific evidence showing that delusions or hallucinations produced by a dysfunctional brain," or even a healthy one for that matter, "can induce the kind of long-term positive changes and psychospiritual transformation that often follow" mystical experiences. Instead, the illusory can actually have, and more often has, a negative effect from the perspective of the experiencer. Beauregard urges that our motivation, after all, should be to detect what the world has to disclose, not to reinforce one's pet theory and insist on the world as conforming to it. A more open mentality would allow us to ask questions such as whether the brain can act as a TV receiver, accepting and translating signals from an outside source, perhaps thereby enabling us to tap into a spiritual dimension. Investigations of this sort could also yield a further question, namely is the brain to be understood as a passive instrument in mystical communication or can it be a sender as well? If so, then the receiver analogy will require modification (Beauregard & O'Leary, 275, 278, 292–93).

The analogy is helpful, too, in another context. Our own death, upon which we sometimes brood, is, as the saying goes, as inevitable as taxes. Some commentators assume that consciousness endures during the lifetime of a higher organism but that the creature's death means the cessation of both body and consciousness. Ducasse responds with this analogy: smashing a radio eliminates any evidence of the program that was being received by it, but this says nothing about the uninterrupted transmission of the same program to other, more functional, receivers. The program persists even if some receivers do not; nor ought one to surmise "that the program was a *product* of the radio." One view of the body/brain-mind relation that we did not cover seems appropriate to introduce here. The double aspect theory is a perspective that deems physicality and mentality as opposite poles on the same axis, with one substance applying to both ends. Taking yet another analogy, the folding of a piece of paper yields two aspects: a ridge on one page and a valley on the other, though neither strictly causes the other. The analogy does not hold, however, if a pencil mark were to be placed on one side, for there is no corresponding mark to be found on the other. Thus only if this elusive substance were to be identified, "of which brain and mind are supposed to [constitute] two 'aspects,' *nothing* can be inferred as to whether the...death of the brain—is or is not automatically matched by the death of the mind." No such substance has as yet been uncovered that could allow us to test whether changes in one aspect has corresponding changes in the other (Ducasse, 61–62, 72–73).

Some authors, such as Stevenson and Haberman, have attempted to extract from the biblical writings an understanding of human nature that resembles this double aspect theory. They tend to refer to humans as psychosomatic unities, having one substance with physical and mental aspects. This conclusion appears strained or at least one-sided to me. A more open reading of biblical authors such as Paul reveals a view that is decidedly dualistic. In 2 Corinthians 5, the apostle Paul teaches a position that Plato would endorse, for the language is similar to the ancient Greek version of the body as a "prison-house of the soul." Paul bemoans that we long to be released from this "earthly tent" that we inhabit, so that we might be freed in order to be with the Lord. We are to be clothed with a new body, that is, there is a we or us that has a body, not that we are equivalent to a psychobody. There are not two aspects here in Paul's thought but two classes of things. Paul states that we are at home in the body, but would prefer to be away from it. No amount of fancy footwork or acrobatics, in my estimation, can prevent the conclusion that Paul uses dualistic language, much as I would rather that he did not.

A few miscellaneous items before we close this section. First, as intimated, not only are mystical meditative states found in non-Western contexts, but since we are on the topic of death, "The content of NDE [Near Death Experiences] and the effects on patients seem similar worldwide, across all cultures and times." Not only death but the specifics leading up to it, even in some detail, are typical and consistent in the human experience. This need not be construed as a statement on religious belief, but, as cardiologist Pim van Lommel concurs, an attestation of clinical findings. NDEs appear to be a fruitful subject for scientific scrutiny. Second, in reference to what Rhea White terms "Exceptional Human Experiences," or EHE, examples other than NDEs can be multiplied, including Carl Gustav Jung's notion of "synchronicity," or "meaningful coincidences." In cases of this type, two or more coincidental occurrences, such as the correspondence of the number of your address with the number on your dry-cleaning pickup ticket and the number on your seat at a sporting event, have no formal causal relationship of which we are aware and in combination are highly improbable. Yet many of us have had this kind of encounter with improbability. The world seems to be structured in such a way that it improbably contains "acausal orderliness." Before we scoff at the claim, keep in mind that these occasions are not foreign to science. Albeit in the quantum world, implying not in our ordinary everyday world of experience but on the subatomic level, events have transpired that point to no well-defined cause. For instance, it is not known which will be the next nucleus to decay in a radioactive substance, and when it does disintegrate, no specific cause can be determined. Thus if quantum events are accepted at the microscale, synchronicity might be accepted at the macro. The difference here, of course, is that one set of physical laws does not cover both micro- and macrocosmic levels; rather, two different sets are required. This is one eventuality which Sir Isaac Newton and his mechanistic system of the universe failed to anticipate (Beauregard & O'Leary, 153; Jung, 456).

Third, if psychoanalytic techniques enable the practitioner to divulge what is occurring subconsciously in the patient based on his or her sometimes neurotic behavior and the reports s/he submits about dreams, what would be the thrust of the objection to telepathy or mind-reading? If these patients give off verbal and nonverbal cues, so too could those whose thoughts might be exposed as more transparent than suspected. There is a difference between reading minds and interpreting cues, but both are signals, even if one form of perception is extra-sensory (ESP). Besides, maybe this is not an impenetrable barrier. Perhaps telepathy is an extension or enhancement of a TOM, whereas clairvoyance would not be if objects come without minds.

Fourth, most creatures are susceptible to biorhythms. They are inescapable. No place in the universe can isolate an organism from conditions that are not constant, for constant conditions, along with vacuums and frictionless surfaces, among other things, are an unattainable ideal. "Even in the deepest, most shielded underground chamber, such factors as tides…or…geomagnetism, would continue to convey some of the periodicity of the cosmos." Electromagnetism, for instance, acts to deflect compass needles and sun storms are one source of this force. Neurophysiology, which runs on electrical signals through chemical neurotransmitters, can be similarly affected. Conceivably, the brain could be sensitive to and "also [respond] to large magnetic disturbances."

A moment ago we broached the topic of the quantum world and now seems appropriate to do the same for the relativistic. Much like Albert Einstein's theory of general relativity, there is a non-linear relationship between gravity and the geometry of space, for one always affects the other. Masses curve space and curved space influences mass trajectories. We can never evade some amount of gravitational attraction. Its pull is unavoidable and it can affect our lives in ways we do not expect. Gravity is pervasive and operates at cosmic scales, as would electromagnetism if the universe were not electrically neutral. Nevertheless, we are bathed in both and they cannot help but influence us. Gravitationally, if objects like the moon are sufficiently massive to affect the tides, then there could be a non-negligible influence on us, given that proportionately we have about the same amount of water as liquid covers the earth, roughly 70%. And electromagnetically, both the universe, in the form of microwave background radiation, and the earth, as a large magnet, have a hum, they make a noise. The former can be detected as light and temperature, the latter as sound waves. We are usually healthier when we keep in synch with the geo-rhythms of our planet.

The fifth and final EHE I wanted to address, and which will lead us into the next segment, concerns reincarnation. If our consciousness is discontinuous due to the lack of infant memories, later memory lapses, being "in dreamless sleep, [under] anesthesia, in [a] coma, or otherwise," is it reasonable to dismiss reincarnation on the same basis? Our failure to recall something in our present life might very well be akin to doing so for an entire past life. This could even extend to personal aptitudes not traceable to our ancestors. If heredity plus environment cannot account for them, then maybe past lives can (Broad, 382; Ducasse, 224–25, 230; Luce, 440–43).

Beyond this life

The sizeable amount of themes treated in our study is needfully cursory; space allows but a mere glance at them, though each could be expanded into a volume in its own right.

Our current theme is death and its aftermath. In the gospel accounts of Jesus' own death, both Matthew and Luke imply that there is an extra-physical component to humans, assuming of course that Jesus is just like us in this respect, which is released upon our demise. Matthew, in somewhat dualistic fashion, has Jesus "giving up his spirit," and Luke has him commending his spirit to God and then "breath[ing] his last." Mark does not go into detail beyond the last breath (Matthew 27:50; Mark 15:37; Luke

23:46). What lies ahead for ordinary mortals "is only the shadowy domain of Sheol, in which pale shades remain like fading carbon copies in a forgotten filing cabinet." In Jewish tradition, humans are understood "as animated bodies not incarnated souls." In essence, we are first and foremost corporeal with a divine ignition, as in the Hebrew depiction, not disembodied souls looking for a physical home, as in the Platonic. Nevertheless, this divine quickening or booting-up-of-the-system is released somewhere apart from the body at death. In Jesus' case, the biblical testimony is not definitive as to whether his spirit went up or down or both for multiple days prior to its being reunited with his resurrected body on earth. Hence it is insufficient merely to hold that God retains the pattern of one's soul in God's memory until such time as the resurrection occurs generally. Contrary to authors such as John Polkinghorne, there does appear in the biblical witness a human component that survives (Polkinghorne (2002), 55, 57; Polkinghorne (2010), 43).

Referring to Sheol as the land of the dead, Crossan & Reed observe that "For around a millennium of its earliest history, Israel did not believe in an afterlife, not in the immortality of the soul and not in the resurrection of the body...It was not a faith ignorant of afterlife possibilities since Egypt was always next door. But...[this was] an exclusively divine prerogative. An afterlife, in other words, was not worth serious discussion." Israel was aware of Egyptian concepts but elected not to accept or draw on them.

There is an idea about what constitutes the point of death, and this view is found across several cultural times and places. A curious OT passage is Ecclesiastes 12:6 which introduces a concept found only once in the scriptures, that of the silver cord. This is a thread that connects one's physical body to one's spiritual or ethereal body. Reports of this cable-like structure state that it can be six feet long, an inch wide, is shiny and elastic, and attaches, among other areas, from the forehead or between the eyes of the physical body to the brainstem region of the spiritual. When humans give up the ghost or spirit, and as the spirit moves ever farther away from the physical body, the cord is stretched, diminishes in thickness and ultimately snaps. The severing of the thread results in the separation of the two bodies, thereby marking the point of death and the release of the spirit. Once there is no longer any attachment, the spirit is allegedly irretrievable, which means the point of no return has been reached. Depictions of these events are sometimes framed in terms of traversing a stream, crossing a river or scaling a wall—the language of overcoming an obstacle such as a boundary or border. Perception from thence forward is from the perspective of the spiritual body. Separation is temporary in NDEs and during sleep, but is permanent at death (Broad, 184–86; Crossan & Reed, 255–56; Ducasse, 9).

Discussions of this sort should not disturb those of a more traditional bent, for we are all unwitting proponents of the paranormal, specifically parapsychology. There are four main branches of psi (short for psychic abilities) pertaining to the power of the mind and the knowledge it can attain or the awareness it can be privy to, two of which we have already mentioned. Telepathy (knowledge of other minds) and clairvoyance (knowledge of other objects) are categories under the umbrella of ESP, and the other two are precognition (knowledge of future events) and telekinesis (the ability to move objects at will without physical contact). We all believe in the last of these. We can raise our arms, for instance, should we have the capacity to do so and come complete with two of them, simply by making the decision. There is nothing physical that mediates between the deciding (mental) and the raising (physical), a topic we entertained when treating the conservation of energy objection to the dualistic theory of the body/brain-mind relation.

Perhaps analogously, some Christians hold to telepathy in the way they deem the holy spirit to operate. They may wish to know the mind of the spirit, in which case they would be reading the spirit's thoughts, or this spirit could function to impart thoughts (as a sender) to those followers eager to do its will (as receivers). What are these impressions but telepathic signals, geared to one mind plumbing the depths of another or one seeking to influence the thoughts of the other by presenting some of its own? Hence there is no need to downplay psi, for we are closer to it than we might expect.

Having left off at the point of the severed cord, we arrive at the stage of death. Recall a prior discussion about the OT book of

Daniel as containing the initial indication of resurrection. There it is trumpeted that many will undergo this change (12:2), while in the NT the event seems to be extended generally to all. Yet there appear to be even earlier intimations of it. One passage in the Psalms could relate the idea that at least some who have fallen asleep shall reawaken, though it is unclear as to whether death is necessarily intended by this falling asleep (17:15). A second passage could very well instruct that the grave or Sheol is not the final resting place for all those who have entered it (49:15). And a third could be taken as ambiguous on this score (88:10–12). Additionally, one interpretation of Job 19:23–9 implies that there is more than simply this life, but the passage could refer instead to the present one.

If reports about the silver cord are accurate, then as we depart from the womb attached to another body through an umbilical cord, so analogously (and we are employing multiple ones) we enter the spiritual world through, ordinarily, a cranial umbilical cord. Thus we go through two gestation periods, the second one far more protracted in that it takes a lifetime to complete. The first is a birthing event which is painful to our mothers as we physically separate from them; the second is a process where the pain of eventual separation is our own.

The NT book of Revelation, which alludes to our as well as the world's ultimate fate, follows Paul in depicting humans in a dualistic way. On three occasions the author(s) use dipartite language and differentiate souls from bodies (6:9, 18:13, 20:4). In the first and third of these examples, souls are portrayed as resting under the heavenly altar or in close proximity to judges on heavenly thrones, awaiting the time when they will be brought back to life, presumably in a renewed substantial yet spiritual sense (the first resurrection). The third is also a passage which refers to a second resurrection which will include the rest of humanity (20:15).

When we covered the topic of dreams, we considered how many of our senses are present in them and tentatively concluded that only taste and olfactory ones are not well represented. We might now ask how many we can expect in a resurrection body. Employing Jesus' own resurrection as a guide, uncertain of course as to how reliable are the accounts, his visual, auditory and tactile senses appear operative. While on the road to Emmaus, in Luke's portrayal, Jesus approaches and converses with two disciples, thereby allowing us to check sight and hearing off our list. The sense of touch also gets the nod as he invites others to touch him in order to confirm for themselves that he is not a ghost. He even ate some fish that he was offered, likely enabling us to include the sense of taste (Luke 24: 13–43). The book of Revelation further teaches that resurrected humans will have access to the fruit from the tree of life, assuming this is something that a form of taste buds could savor (22:2). The question is whether the remaining sense of smell will also function in the hereafter. The OT indicates, likely metaphorically, that God enjoys the aroma of burnt offerings (for who can resist a barbecue?), but it is not definitive as to whether this pleasure will extend to resurrected bodies, since not even Jesus is reported as possessing it.

Questions of this type yield another concern, namely the qualitative one of whether, should the afterlife not hold olfactory cues, the experience will somehow be impoverished for us? Moreover, if there are neither post-death genders (was Jesus' resurrection body still male?) nor marriages, then would the absence of eros also diminish the quality of life? When reflecting on the similarities and differences of living in these two eras, here is yet another item to consider. If there is to be a separation of sheep and goats, meaning a distinction of insiders and outsiders when, using biblical imagery, the gates of the kingdom are shut (as with the doors of Noah's mythical ark) and when there will be weeping and gnashing of teeth as a result, what kind and amount of comfort could God bestow on insiders who regrettably have loved ones on the outside? Would this elicit weeping on the inside as well? The book of Revelation states that God will wipe away every tear from our eyes (21:4). Is this where some of those tears will be produced?

Now that we have looked at the sweep of the human life cycle and even beyond, we can ask if we are any closer to identifying our nature. Maybe this will become clearer with the following case studies. Quite possibly, we might catch a glimpse of who we are as we examine what we do, the kind of behavior we engage in. To this end, data is not difficult to find; we do not need to search very far to obtain it.

PART FIVE

Case Studies

NOW THAT WE HAVE SEVERAL THEORETICAL THEMES about human nature under our belts, this section of our study examines two documentary films and asks how they might inform us as to our natures. They are, in order of release dates, "Expelled" and "Religulous." The former deals with the cold reception that faculty in education face when speaking to the issue, let alone holding the position, of whether the topic of Intelligent Design (ID) should be raised in these settings, while the latter investigates what passes for religion, specifically the three Abrahamic traditions of, chronologically, Judaism, Christianity and Islam. But I wanted to commence with a warm-up exercise.

I penned the following meditation as I observed the workings of a US political campaign. All candidates therein made an appeal to the perceived foundational ideals of the nation, framed in such a way that they would appear to be exemplifying the party position on major issues. I decided to offer some reflections of my own and tease out what the deliberations might reveal about human nature. In order to accomplish this, I present a simplistic though useful distinction between conservative and liberal.

A conservative mentality as portrayed by Republican sensibilities would favor the stance that liberty should be promoted at the expense of equality. Persons should be free to pursue and exercise individual rights and freedoms even if neighbors are left behind. On the other side of the political spectrum lies the liberal viewpoint, the proponents of which, at least in theory, are the Democrats, whose emphasis is on equality even if it is at the expense of liberty. The welfare, or employing the less politically-charged term well-being, of neighbor is upheld even if it means individual liberties are compromised. The debate rages as to which of these two should be adopted as social policy, for what is at stake lies at the very heart of American self-perception, namely what does it mean to be an American and to live out its dream?

Add to this the self-identity that the US is a Christian nation and the following difficulty arises. The trouble is that both sides, conservative and liberal, understand themselves as representing the true biblical position. As the gospel writers depict it, Jesus as Messiah has set us free. For conservatives, this means freedom to live as we please under God, to take life on in an unrestrained way as a follower of Christ, for this is the essence of liberty. On the contrary, for liberals, the previous notion has its faults, for it does not go far enough. Liberals ask what precisely it is that Christ has set us free *for*. Taking the gospel example to heart, God through Jesus has bestowed freedom upon us for the expressed purpose of serving. Jesus came to serve and now asks us to do likewise, thereby following his example. In the first instance, we are understood as having been unshackled and free to take the reins of life in God's creation. In the second, we have replaced one set of fetters with another—we are unfettered when it comes to the old nature and fettered when it comes to the new. To follow Jesus means to be in service to God and neighbor. Conservatives tend to de-emphasize the second of these, for others are sometimes deemed as impediments to personal advantage.

Coming from a land of socialized medicine as I do (Canada), the health needs, though not the wants, of the other are attended to as a guarantee, not based on merit but as to what is deserving as a basic human right to which all are entitled. The wants and desires of self are subordinated to the rights of others, and this is what poses a threat to conservatives. For them, individual liberties are sacrificed for the sake of the common good in liberal policy. The conservative counterargument is that liberty guarantees the potential for people to do good. For liberals, that very good itself is guaranteed (at least in theory). Conservative ideals can bring out the very best as well as the very worst in human nature; liberal ones ensure that there will be at least a minimum of good.

The political strategy on the part of conservatives is to villainize socialism as a policy that limits freedom and undermines the liberty upon which the American nation was founded. The difficulty, however, is that all humans tend to be self-serving as opposed to other-serving and are disinclined to care for neighbors on their own. Sometimes they need a push. Observation of Wall Street methods of operation amply reveals and makes abundantly clear that attention to the needs of others is not at the forefront of unbridled self-interest. Conservatism tends to put self first; liberalism tends to elevate others and is also more inclined to celebrate individual differences. Conservatives stop at individual liberties, making America a nation of individuals. Liberals extend the vision beyond individual rights and freedoms to include the other, making America a nation of level playing fields for the expression of individual abilities. Well-being is more evenly distributed in a liberal setting and less so in a conservative environment.

In my estimation, humans are naturally inclined to be self-serving, and in this vein the US is naturally individualistic and hence conservative. Both the terms liberty and equal occur in the Declaration of Independence, but which is to take precedence? In an individualistic society, likely the former. The natural inclination is therefore to be conservative; one must work at being liberal. People do not need to be taught to self-serve. Conservatism leaves people to fend for themselves; liberalism enables them at least to live and fight another day.

A handy way of diagnosing which camp we fall into is our reaction to an unemployment situation. If we see unemployment as a failure on the part of the individuals themselves, then they carry the weight of the blame, for it is their responsibility. This is a conservative position. Alternatively, if we view it as systemic, where the system has failed those same individuals, then all of us bear the responsibility. This is a liberal stance. The actual situation might probably lie somewhere in between and may differ on a case by case basis.

In an interesting twist, conservatives sometimes have poor expressions of Christianity yet claim to march under its banner. An example would be in their emphasizing untrammeled free market economics even to the detriment of the impoverished and disenfranchised for whom Jesus also cared. Liberals, on the other hand, sometimes have commendable expressions of Christianity, such as social programs for the needy, though may not intentionally reside under its banner but under, say, humanistic empathy and concern. Despite the conservative insistence on liberty at the expense of equality and a de-emphasis on Darwin, whose ideas they tend to villainize, they are sometimes being more Darwinian that they realize by allowing the socio-politico-economic climate to decide who survives the market competition. Business is business, after all. Conversely, in spite of the liberal insistence on equality at the expense of liberty and a stress on Darwin, they are sometimes being less Darwinian than they realize in not allowing market forces, out of charity, to completely dictate who survives the competition. There are times, in their view, when it is more important to assist the downtrodden and maladapted, as in the physically and mentally challenged. Which of the two, we might ask, is being more Christian?

Having broken a sweat in our initial exercise, we move on to the lead up to our two films.

Nothing gets the blood boiling like stories of injustice, and nobody provides them like media sensation. The media generates reports with a punch by appealing to our sense of justice and fair play. News magazine type programs like *60 Minutes* are able to capture our attention by delivering items with a "something needs to be done about this" edge to them. For me, their steady diet of editorial pieces make my own Sunday evening meal more difficult to digest.

When presented with a particularly juicy piece on injustice, I can imagine my blood pressure rising and notice myself adopting emotive postures I normally would not. Lynching, a term derived from the surname of the one responsible for putting forward this innovation, "is too good for them," I catch myself thinking, no doubt the intended response on the part of the show's producers. After you collect and compose yourself, you are in a more proper frame of mind with which to assess the cases presented. Maybe there are other sides to the story not reported, implying that there is more work to be done than that covered by the journalists in the time allotted.

I recall interacting with an associate about biblical authors and urged that they wrote what they did because each had an agenda—a message that each wanted to convey even if it meant taking liberties with the accounts. Like journalists, they have a story to tell, only part of which accurately reflects the actual state of affairs. Invariably, we all operate this way and, as intimated in the section on the nature of science, come with our own perspectives which color, distort and put a certain flavor on our perception of reality. Sadly, as there are as many interpretations of an event as there are interpreters, the actual occurrence is forever beyond our reach. Kant began to alert us to the mind's involvement in constructing what the senses experience, so much so that we can only ever grasp our "take" on what we observe and not the "thing in itself." We may attempt to be objective, but we can never disinfect or decontaminate our subjectivity from it. My colleague misunderstood this appraisal of mine as applied to the biblical authors' portrayal of Jesus as amounting to intentional subterfuge on their part, unbecoming for a writer of sacred text, so it must be a mistaken view. Agendas could only be viewed as negative and something to which biblical authors would not be susceptible. Regrettably, my interlocutor completely missed his own agenda in insisting that the holy spirit would assist in eluding this pitfall, ensuring that an unadulterated account would result. What results, however, is a sanitized version.

For this reason I wanted to examine two films as media products, complete with their own agendas. But first another item on an autobiographical note before we proceed.

A word about the college at which I most recently taught. It is a small Jesuit school, where it and I had a beneficial working relationship. We left each other to our own devices. I am grateful that there were no watchdogs looking over my shoulder to ensure that the content of my didactic activities conforms to the dictates of tradition. I was not called upon to jump through certain theological hoops; they have too much respect for the academic process. "If ideas have merit, they will become readily apparent in due course and will stand the test of time," seems to be their policy. I heartily agree. We need not fear ideas, whether old or new, though the uses to which they are put require careful scrutiny. Some of my students wished that I would have been more traditional, for then they would have known what to expect. But education is not about expectation. Education would be an exercise in recall if we knew what to expect. When assumptions are dismantled, presuppositions unmasked and foundations shaken, then education has a chance to occur. Some might become unnerved, even threatened by this, while other fertile minds revel in the growth potential. One can at times feel the "ahas" percolating in the room. Educators are to "comfort the afflicted and afflict the comfortable." The former is the task of a counsellor, the second a critic; teachers engage in both.

Not all schools are like this one, however. Other confessional schools require their faculty to sign a statement of faith so as to ensure that they do not stray from the institution's "agreed creed." In some colleges these statements need to be signed on an annual basis, largely to allay the fears of the administration and board of trustees. They do not want anyone veering from the accepted path. If any faculty were to entertain the thought of a change to a rival denominational affiliation, they are shown the college's door. I know such tertiary level educational institutions exist, for I interviewed at some.

What irks me is the denial of that which institutions of higher learning are allegedly built upon, namely academic freedom—the latitude to go where the research results take you, even if it means a modification of convictions. This is what education sometimes does to those who reflect on what they learn: their viewpoint changes, their perspective shifts and they may even adopt a different theoretical allegiance. Why would anyone attempt to hinder this? Yet it happens. Most likely indicative of the desire for control. And the unfortunate truth is that it also occurs in educational settings that ought to know better. There is not as much academic freedom in purportedly non-confessional halls of learning as was formerly supposed. And if this is the case, then they actually do have an agenda, a mindset to uphold, a philosophical position to defend, which makes them confessional schools after all.

This is the theme of the film "Expelled: No intelligence allowed."

Expelled

Our first film is hosted by TV personality Ben Stein, who like me has similar misgivings when it comes to a certain type of injustice. At some tertiary level educational institutions, that is colleges and universities, which one would expect to be bastions of free thinking, there occurs a suppression of thought. They are hailed as sanctuaries for open inquiry even if surrounded by anti-intellectualism, but Stein uncovers a flaw. There has, on the one hand, been a breach in these defenses and, on the other, a wall of another kind has been erected in its place.

The issue revolves around creationism's more sophisticated relative, namely Intelligent Design (ID), though creationism perceives ID as a sellout—an accommodation to the evolutionary enemy. I will elucidate its perspective momentarily. In the meantime, what Stein finds reprehensible is that in the course of otherwise reputable scientific investigations, respected scientists have entertained the idea that ID might be considered a legitimate alternative explanation of how life and life forms developed to the standard Darwinian evolutionary program. In going public with ID views, these faculty members have paid a high price for their insubordination and have found that dissidents are unwelcome. Several have lost their employment for stated reasons that evade the topic of ID.

The message being sent is that tolerance of ID is injurious to one's career, so it cannot be tolerated. Evidently academic freedom only goes so far. Voices that speak freely of ID, whether approvingly or merely in passing, are muffled. Even if scientists of any stripe were to express their concern that Darwin's theory could benefit from renovation, liberty does not extend to freedom of speech, this documentary contends, when it comes to those seeking to refine the Darwinian gospel. Criticizing Darwin, it seems, is considered heretical and could only stem from the lunatic fringe, particularly the creationist camp, so it must be censored. Science must defend the party line and thus censure offenders by silencing them.

Yet bucking the trend is precisely what scientists, in often forgotten historical twists, have been known to do. Lord Kelvin advised Max Planck at the turn of the twentieth century not to pursue a career in physics because there was little left to achieve outside of extending already existing numerical data to additional decimal places. Only two phenomena, the photoelectric effect and blackbody radiation, stood in the way of a complete Newtonian picture of the universe, and how long could it take for those two to conform?, asked Kelvin and his cohorts rhetorically. These last holdouts were simply delaying the inevitable. But they proved to be the foundation for two revolutions in physics, relativity and quantum theories. The prevailing Newtonian paradigm had become supplanted. This is one way that science advances, by questioning the ruling paradigm. ID is in that very tradition. Newton was not the final word; perhaps so too with Darwin.

ID is seen as a threat to the safety and security of natural explanation. If the door is opened to religious causation, so the argument runs, then science is undermined. Hardly an airtight rationale. The trouble is that even a luminary like Newton was devoted both to his religion and his science, the former inspiring the latter. Newton allowed his religion to inform his science. A consummate scientist, perhaps the greatest in history, could very well be shown the academic door in the current scholarly scene. Indeed, in the contemporary climate he would have been relieved of duties, expelled, were he on faculty. He believed that the study of the heavens and the world of nature is a study of the handiwork of God. With this mindset, there is no place for him in the academy. To reiterate, freedom of thought does not always translate into freedom of speech.

While for some the path to progress is to downplay, even eradicate, religion, I have never understood how learning more about the world *must* detract from religion. Newton did not see it this way and he has modern day peers. There is a difference between open inquiry and insisting that the findings of such inquiry overturn a transcendental component. ID concludes that something more than simply material reality is at play in the history of life; critics not only reject this claim, which is their right, but demand that it cannot take place. This is overstepping their bounds; in fact it is not science. Contrary to creationists (and this was the correct ruling by Judge William Overton in the Arkansas trial in 1982) who do not conduct science and therefore their views should not be taught in science class because their approach is not scientific, IDers actually do carry out science and come to a certain conclusion about it. Some scientists in the other camp have concluded not that the science of the ID crowd is faulty but that their conclusions are overruled from the outset. IDers are free to speculate about their findings as critics are of theirs, but to reject ID speculation on the basis of religious content is beside the point. The critics are being, either deliberately or unwittingly, religious in their insistence on an anti-transcendental bias. The time has come, the film declares, for science to be self-critical and recognize its own blinders.

Toward the end of the documentary, in an interview with everyone's favorite rabid atheist, Richard Dawkins, Stein gets Dawkins to admit that ID may have its merits. Dawkins maintains that an intriguing possibility concerns a race of technologically sophisticated extraterrestrials which could have seeded the earth with life. Well behold, this is a form of intelligent design, even if it is of the lower case variety. Admittedly, the ultimate question of the upper case divinity version is pushed back to another stage, but the point remains that the origin of life on our planet according to this scenario has an outside source. It might very well be natural, but it is alien to our earth and it amounts to a design by an intelligence. So Dawkins cannot have it both ways; either life arose all by itself or it had help. In this instance, Dawkins unknowingly agrees with the ID perspective.

Having said all this, I am no friend of ID. But that is not the issue, for I find myself supporting them when the scientific inquisition comes to town, for then we have a common cause. In the face of such opposition, they could use some assistance. But I will allow neither science nor religion to tell me what metaphysical position to take. Here now is a fuller exposition of the ID platform.

IDers suggest that God's hand (or finger as being the most likely body part) is implicit in nature given the complexity of the world. This type of argumentation, known as teleological and having to do with purpose, has a long history and an impressive pedigree, a line beginning with Aristotle and working through Thomas Aquinas (c.1225–1274) and William Paley (1743–1805). Paley popularized this strategy by proposing the following. If someone were to stumble upon a watch as s/he is walking along a path, the assumption s/he would make is that because timepieces are precision instruments, they could not arise simply through natural forces alone. We would be compelled to declare that what we have before us is a human artifact. (This is of course sidestepping the observation that since humans themselves are products of nature, so too, at least indirectly, are their own artifacts. Nature produces more nature in the act of reproduction and in the creative process. But this is perhaps to equivocate on the meanings of the terms nature and natural.) In like manner, and this is still Paley talking here, as we would point to a watchmaker as responsible for the watch, so too would we be struck by the intricacy of the human eye in particular and the world in general and in turn be driven to conclude that there must be an eye-maker as well as a world-maker. And who else could fit the bill but the divinity?

Paley had, from our vantage point, an unexpected supporter in the figure of Charles Darwin. The early Darwin from before the voyage of the Beagle was an ID enthusiast. Sailing around the world and plying his naturalist trade, however, cured Darwin of his previous teleological leanings, although the process was a slow one. Darwin eventually surmised that even the eye can be explained solely on the basis of natural causes, and recent biochemistry has reinforced this claim. Contemporary cosmology further asserts that the world itself is the result of a long series of natural astrophysical occurrences. If God is anywhere in all of this, so the reasoning goes, it must be at the very beginning which is inaccessible to science. This is where both philosophy and religion get to chime in with their own views on a somewhat equal footing.

IDers are more in tune with current scientific advancements and are more willing to accept the findings of recent developments than their creationist cousins. As such they often hold to both an old earth and an old age for the universe. In essentials, they agree with the contemporary understanding of the history of the cosmos and the evolution of life on planet earth. Where they differ is in the details of the mechanism. They see the Darwinian description of the ability of natural selection to cross classification boundaries from genus and above as stretched. Over the course of time a new species can be produced but not likely a new genus, family or order, and definitely not new classes or phyla (Behe, 218). Thus IDers maintain that under its own steam evolution is ill-equipped and lacks the wherewithal to accomplish what Darwinists claim it does. Despite adhering to common descent, for IDers the professionals (read divinity) need to be called in to arrange what natural selection cannot. And whereas some religious philosophers contend that evolution accounts for life's general brush strokes and the divinity attends to the details (recall it is said that this is actually where the devil resides!), IDers opt for the reverse, namely that the Grand Designer crafts the general plan and nature fills in the blanks, a position similar to Darwin's own. Only in one of these ways could evolution accomplish what it has in the time allotted, if God were to be its driving force in expediting the onset of adaptive variations. Details might not be complex, such as the amount of hair we have, but where there is complexity it stems from the complexity of God.

One weapon in the ID arsenal is the disclosure of several amazing coincidences in the physical world. The notion that the universe is exquisitely fine-tuned is known as the anthropic principle, where the way the world is set up leads to the arrival of humans. The cosmic mystery is why everything is ordered just right seemingly in preparation for our appearance. The conviction on the part of IDers is that God designed the world this way; chance cannot explain the formidable improbability. Only God could pull it off, conscripting the natural world to combine in the precise way so as to culminate in authors for Blue Dolphin. No one other than God could cut the world that finely, all of which points to divine fingerprints. Who could fail to be convinced by such impeccable reasoning?

I for one am not. I suppose that amazement lies in the eye of the beholder. It might boil down to a perception issue. Try another coincidence on for size. In the early history of baseball, a decision was reached as to the distance between bases. What would be the optimum amount, it was asked, for a base runner to travel for purposes of the benefit of the game, especially maximizing the sport's impact and entertainment value. Too short and the scores would resemble future basketball results. Too far and home runs could account for most of the scoring. Ninety feet, thirty yards, was settled on as just the right amount. It constitutes what would be close plays for a batter to reach first base before fielders could deliver the ball there, or for a catcher to throw out a base stealer running from first to second on a diamond-shaped infield. No other distance will do. What an amazing idea to hit upon. And it has worked ever since; in the course of time, faster runners keeping pace with stronger throwing arms. It's the stuff of genius. Does this mean the divinity must have played a role in the proceedings? Or is there intelligence in trial and error if not the merely haphazard? Amazement always derives from hindsight, after the fact. And the fact is that what remains of a process after the dust settles can appear amazing to those who have eyes to see in that particular way. But not all interpret it in such a way that insists on the collusion of otherworldly forces, despite appearances. If one does not need to resort to an appeal to God for the dimensions of baseball, perhaps something similar can be applied to cosmic dimensions. Besides, my hope is that God is more of a football fan.

Thus neither religious nor scientific orthodoxy should stand in the way of progress. I can accede to this regardless of my own personal inclinations on Darwin and ID. I am defending those procedurally with whom I disagree theoretically. So to be thorough, I would be remiss if I did not also mention a certain inadequacy. For those finding themselves in this fraternity, they attach a certain type and amount of religious fervor and zeal to their position. This after all are the contours one expects from conservative leanings. I speak of course, it need hardly be stated, of staunch evolutionary theory. Were you anticipating ID instead?

My misgivings stem from a lack of certitude in this scientific endeavor beyond the level of conjecture. One expects that scientific pronouncements have a greater trustworthiness than ordinary subjects of speculation. But the rigidity of evolutionary doctrine suggests otherwise. This is a terrible thing to do to a believer in evolution like myself. I thought there would be more bedrock than interpretation to evolutionary history, yet such is not the case. This makes evolution more of a belief-oriented enterprise similar to a religious stance. I expected more from those who have been entrusted with our history, how we got here. Now it becomes suggestive that this is not so obvious. Where once I could see, now I am blind (to reverse the line from "Amazing Grace"). Some of the difficulties surface in the following ways.

The scenario of beneficial variations being preserved by (the alleged force of) natural selection and detrimental ones being pitilessly weeded out, is a heroic one. But too many other factors can get in the way, about which natural selection has little to say. Adaptive variations can accumulate, until, all of a sudden Bambi meets Godzilla and perhaps a potentially promising set of traits licks the forest floor. Or out of the blue, a volcano spews out lava destroying, either directly or indirectly, most plant and animal forms. There is no gene for what to do in such a situation. No variation, no matter how adaptive, can withstand such an onslaught. Lava wins; organisms, at least for the time being, lose. The most fit do not necessarily survive or leave the most offspring. Adaptations are no guarantee for longevity as a species. Natural selection displays no foresight in how to prepare for an uncertain future, but operates only for the moment. Romantic tales of champions of fitness are histories that require revision. Our line might not have been the fittest in terms of adaptive variations but the most opportunistic in terms of being at the right place at the right time. There is no gene for that either.

The problem amounts to the scourge of fundamentalism. Here are some more scientific examples.

First, materialism. Those who stridently declare that they are materialistic, perhaps thereby seeking to communicate that they have come of scientific age in the twenty-first century, and that all others should also follow suit else be at risk of seeming immature, are unaware of their error. For theirs is a philosophical and not a scientific stance. Science is not in a position to assert that all of reality is material in nature, since this is a metaphysical statement, unbecoming for a scientist. Nor can all of reality ever submit to laboratory conditions. Simply wearing a labcoat does not allow one to bear the mantle of a philosopher. More work is required.

158   *Who Do We Think We Are?*

Further, it is not even clear as to what matter actually is. Staking one's claim on the unknown sounds religious to me. I wish I had their faith!

Second, dark matter and dark energy. It seems that only about 10% of the matter required in galaxies to keep them intact is actually observed. Brown dwarf stars, mini stars that do not shine in the visible part of the spectrum for us, though can be detected using instruments that capture infrared light, could very well make up some of the difference. Theories propose that the vast majority of the material in the universe is comprised of dark matter (23%) and dark energy (72%) and therefore this is the explanation as to what holds galaxies together. Cosmologists have been known to announce that dark matter and energy are unobservable but they must be there. This proposal sounds similar to John Wisdom's story about the invisible gardener, where an unseen gardener is posited because a plot of ground appears to be tended, despite the analysis being interpreted as strained. How is this an improvement over religious statements? Are we insisting on invisible gardeners? And are IDers guilty of the same? If this passes for science, then science should not be so critical of religion.

Lastly, evolutionary psychology. This sub-discipline discusses how natural selection acted on behavior in the long evolutionary history of humans. Behavior, though, is not something that leaves fossilized remains as such. Certain practices, admittedly, leave behind artifacts, especially those involved in ritual, art and warfare, but intentionality as a non-scientific category does not. What early humans must have been like in order for us to comment on their perception of the world, for instance, is not history. Nor can we be certain about their motivations. One diagnosis is that early humans may have reasoned that one should not venture out of a cave because evil spirits are lurking about, when in fact there was a lion on the prowl and could have killed this early religionist. The conclusion is then drawn that religious beliefs are adaptive since they aid in survival. A rival diagnosis, however, can be that a caveman could have believed that, using religious portents, the spirits were in fact emboldening him for success in the hunt, whereupon he ventures out and is killed by that same lion. In this case religion would make cavemen less fit. Thus religious observance becomes ambiguous. Whether religion is or is not fit then depends on one's interpretation. We have difficulty uncovering the intentions of even contemporary individuals; how much more so for those on whom we cannot experiment?

These examples can be multiplied, but here is a final one: the episode of a Bahamas-Atlantis connection. In the first half of the twentieth century, renowned US prognosticator Edgar Cayce, in one of his supine trances, prophesied that in 1968 a discovery would be made off-shore from a small Bahamian island of a structure indicative of an ancient civilization that could be linked to the lost city of Atlantis. Sure enough, Cayce had the time and place correct, though the Atlantean provenance is of course another matter of interpretation.

Submerged beneath the waves is a row of stones that some investigators interpret as a product of the sea and others as intentional remains. The stones appear to be too squared and spaced out too orderly to be natural, and there are those who contend that they must therefore be architectural, that is cultural artifacts. The point is not to demand one interpretation over another, but that archaeological attempts at research are stymied and met with opprobrium. To devote resources to tackle the question seriously spells certain disaster for one's career because it is simply not scholarly. To fathom (pun intended) that there could be an explanation other than the natural one is to invite injury to one's credibility and place one's academic status on the fringe. Regardless of what side of the debate one favors, there are parallels here with ID.

It is astounding to me that the traditional, old school even fundamentalist mentality which the scientific community vilifies in others is found also on its own doorstep. This body seems to have emotional investment (read subjectivity) in not only what is to count as data but also as a research program, for they are to be sought and obtained in the standard materialist vein. Their philosophical commitment is such that technologically sophisticated ancient cultures cannot exist since the accepted anthropological trajectory tells us so. But it all depends on what you decide will inform you. To suggest that there might be a rival account is to be subject to derision, for that would upset everything. The conservative voice, found even in academe, insists on maintaining the status quo. Those who would threaten it will be relegated to the ranks of the unwashed.

Thus hypotheses of this sort are in the same heretical boat as ID, for they militate against the official story. And dissenters are made pariahs. Consequently, fundamentalism is unbecoming wherever it is found, both in biblical literalism and scientific materialism. On the one hand, no biblical author, as we have seen, is without an agenda and this makes the enterprise anything but dispassionate. And on the other, materialism is a belief structure that may or may not reflect reality. The bottom line is that there is a seepage of human elements into any undertaking, no matter what the discipline, that contaminates any alleged objectivity. This is inevitable, and the sooner we come to terms with it the better.

Moral of the story: all schools are confessional, since no exposure to education/learning is without perspective on the parts of both teacher and student. This brings us to our second film.

Religulous

If, as our first film made plain, part of human nature is to be territorial, to institute protectionist policies or engage in turf wars concerning what to think, our next film concentrates on challenges about what to believe. By way of the title "Religulous," host Bill Maher is intending to convey his view that religion is ridiculous, using a shorthand version of those two terms. Maher, himself a product of a Jewish mother and Catholic father, lampoons all three Abrahamic traditions as illegitimate human expressions. His indictment revolves around the perspective that their beliefs are irrational and practices immoral. For instance, he judges it nonsensical to believe in both the Garden of Eden and its talking snake, and unethical to participate in the bigotry, misogyny, homophobia and violence that he perceives marks religious history. For Maher, the root cause of humanity's ills is its religious underpinnings. Better, he asserts, to honestly promote doubt over the religious selling of certainty or "blessed assurance."

As the documentary comes to a close, Maher submits these concluding statements: "Religion must die for mankind to live… Faith means making a virtue out of not thinking. It's nothing to brag about. And those who preach faith…and elevate it are our intellectual slaveholders, keeping mankind in a bondage to fantasy." Religion "spawns lunacy" and "is dangerous because it allows human beings who don't have all the answers to think that they do." The hallmark of religion is "arrogant certitude," something to be avoided at all costs. If we were part of an association that behaves the way religions do, Maher declares, we "would resign in protest. To do otherwise is to be an enabler, a mafia wife." He ends his diatribe with this chilling ultimatum: "Grow up or die."

Harsh words. Allow me to commence with a response by admitting that much of what Maher reports is true. The trouble is that those he has in his cross-hairs are usually fundamentalists and they are an easy target. Most thoughtful religionists would agree with his assessment. A major flaw in his report is that he does not consult enough of them. Permit me to redress the imbalance.

Initially, some might declare that Maher's approach is not to understand but rather to ridicule, as the title of his film suggests. Yet he does attempt to spur believers on to recognize and question their own assumptions. Should they fail to do so, then he will poke fun at their presumption. There are times, though, when he goes beyond this and actually seeks to belittle. Plus, his method of operation is not as a journalist but as a debater with a position to defend. And his invective is presented at the sophomoric level—nothing that a bit more study on his part could not remedy, provided he is sufficiently open-minded to do so.

As mentioned, he does not extensively interview scholars in the field and as such his sample is not representative. Those like the physicist from the Vatican Observatory, Fr. George Coyne, Ph.D., acquitted themselves fairly well, though there were also errors in his specifics. (Hebrew writing is not traced back, as he outlines, to 2000 BCE, for they needed to have an alphabet first, and that did not occur according to some views until roughly 800 BCE.) Others, such as the scientist Francis Collins, were not as well versed in the religious details, and only some, like the disillusioned priest outside the Vatican, could laugh at the follies and idiosyncrasies of their own religious tradition.

To cite one example of Maher's superficial reasoning, he understands the portrayal of the OT God, who jealously insists on the exclusive allegiance of his followers, as petty. He does not appreciate, however, that this reaction is precisely what we value in others. We expect Americans to pledge allegiance to the flag; only non-patriots refuse to comply. We further expect fidelity on the part of marriage partners, to forsake all others. God simply asks us to do likewise. Few would claim that these acts are petty. On the contrary, they would be upheld as the height of rectitude. And so they should. Perhaps we can take our cues from God in this regard.

Anti-religionists are not the only ones with these types of misgivings. There are many religionists who have the same or similar objections as Maher to the fundamentalist end of the spectrum, for their own wider camp is no doubt populated with them as well. Had Maher concentrated on the more thoughtful elements of religions, then his documentary would have been much different, or he might not have had a film at all. The media knows that dialogue is not newsworthy; heated debate is what sells. As it stands, the film comes across as one-sided in urging that religion in all its forms is inappropriate, for the exercise of faith amounts to an abdication of reason. But for those who recognize that faith can and needs to be refined in the crucible of rationality, the film is more of a straw man; they are aware that such extreme approaches are inadequate.

In actuality, faith is not so easily eradicated, nor can humans function without it. We have faith that the world will continue to operate in an orderly fashion, even though we have no access to the future. The future cannot be observed, hence it cannot be submitted as evidence. And this makes any appeal to the future unscientific. One can project models into the future, but this does not make it a subject of scientific scrutiny. This just tells us that we have simply placed our confidence, our faith, in our models, that they will be reliable in the days ahead. Yet discourse about the future is metaphysics, not physics. It will not conform to laboratory conditions, nor consequently can data be drawn from it. When the move is made from physics to metaphysics, there is a departure from science to speculation, from observation to opinion. There can be rigor in philosophy as well as science, but when labcoats don the mantle of the philosopher, they would first need to obtain the requisite tutelage, something that labs by themselves do not provide. (Try reading Martin Heidegger without formal training.)

In the social sciences, as we saw, Karl Marx believed that his brand of atheism would usher in a utopia of social salvation. This is a secular faith. Freud also had the view that an unobservable unconscious exists. That could also be considered a religious illusion. Particle physicists believe that protons and neutrons are composed of three quarks each, even though quarks are impossible to isolate. Some cosmologists believe that our universe is but one of many others, even though these are inaccessible. Still other physicists hold that ultimate reality consists of strings and superstrings. The mathematics may point to them but that is not the same as observation. Besides, abstract mathematics does not make its subject matter concrete. So why the attitude when it comes to faith? It seems that even scientists never leave home without it. This calls for humility on both sides.

Part of what it means to be an image-bearer is the ability to reflect on meaning and purpose, individually about oneself and collectively about the world. The discipline of theology is critical reflection on the belief and life of a religious community and its tradition. Maher accuses religionists of not sufficiently engaging in critical reflection. If they did, there would be fewer abuses of various types in religious circles. Well then, let's do some of this critical work.

Another human tendency is to strategize. In Catholicism, the chain of command is in the shape of a pyramid. At the top is the Pope, who is the preeminent figure in the ecclesiastical hierarchy. At the base of the pyramid and on opposite sides lie the Bible and Church tradition, the latter comprised, for instance, of the Church councils and their literary output, such as its creeds. The Pope is elevated in status, while the other two are approximately equivalent in authority but subordinate to the Pope.

Once the Protestants broke away from the Catholics, they had to decide what their own chain of command would be, for the accountability structure could no longer resemble the Catholic format. At the time of the Reformation, and increasingly so since then with the proliferation and fragmentation of denominations, there is no globally agreed upon human at the top to whom appeal might be made. Church tradition, too, goes by the wayside since each Protestant denomination has a different one. The old pyramid is gone. This leaves the Bible.

Protestants tend to concur that much stock should be placed in the Bible as the lone surviving item from the pyramid. In evangelical circles, the Bible becomes the final arbiter in all matters of faith and practice and the source for settling all theological disputes. In fact, this is what constitutes evangelicalism—it is a statement on the authority accorded to this sacred text, and the Christian life flows from the instructions given in its pages. Whatever else evangelicalism entails, at base it means at least that. For some the scriptures become so important that the status of inerrancy or infallibility is conferred upon them. Immediately there is a problem, however. To cite one example, the gospel of Mark terminates at chapter 16 and verse 8, implying that verses 9 through 20 are a later addition. Since they are not in the original, should the longer ending be treated as scripture, let alone inerrant and infallible, by evangelicals? If so, should all later editing, when discovered, be treated similarly? If not, then why does it remain as part of the text? The reality on the ground appears to militate against their expressed policy.

Martin Luther, at the time of the Protestant Reformation, added another authoritative element. He viewed personal conviction as roughly on a par with the biblical witness, and he understood the spirit to work through both the written word and one's conscience, whose dictates generate a spiritual individualism of sorts. The criticism, of course, is that this situation of our accountability hinging on God and conscience potentially leaves us with a relativism of doctrine and practice. The bottom line is that humans decide who and what will assume the role of their authority structure in important matters, religious and otherwise. Perhaps the fervor with which these commitments are sometimes adhered to could benefit from re-evaluation. Or else Bill will get in your grill.

Where I think Maher is right on target, though, is in his appraisal of avenues of divine revelation in relation to this authority we are ascribing to various items. Here is a biblical example.

The apostle Paul had a life-transforming experience, so the accounts have it, on the road to Damascus. The book of Acts chapter 9 informs us as to how Paul (then Saul) interpreted a bright dazzling light and a loud thunderous voice as the risen Jesus, whose disciples Paul was then persecuting. This encounter moved him to an about-face in terms of his mission. The issue for us is how the experience was perceived. When the book of Acts recounts the event elsewhere, the details are modified each time. In 9:7 his travelling companions "heard the sound but did not see anyone;" in 22:6 "they saw the light but did not understand the voice;" and in 26:12 all those present "fell to the ground because of the light." So which is it? Did the event for those with Paul include a video or audio component or both? The least we can say about the account is that Paul's experience was not entirely private, yet at the same time it was not identical for all observers. In essence, they and we must rely on the veracity of Paul's testimony, together with the way the author (presumed to be the same as for Luke's gospel) has depicted the proceedings.

This can be applied to other instances as well. There were no witnesses for Moses and the burning bush or his meeting with God on Mount Sinai when he received the stone Law tablets. No other humans were with Jesus when he was tempted by the devil in the desert. Jesus' resurrected body was seen only by "the brothers," despite their having been five hundred in number at once (1 Corinthians 15:6). And so on. (Indeed, the resurrection was not an objective occurrence, open publicly to anyone, for it was revealed only to a select few, according to Acts 10:40-41.) At least Paul had witnesses, albeit they had a different encounter. The point is that we all need to trust the legitimacy of someone else's version of events. Christianity is based on this limitation on our part; we were not privy to crucial religious-scale events. Regrettably, anyone can say "I had a revelation, so follow me," or "God told me to tell you…" The foundations of the faith are not public; someone else's word must be taken as gospel truth. The difficulty is that this confidence can lead, and often has led, to abuses of power. There are those who have used their authority to take advantage of others. If the spirit does not assist in making the distinction between authentic and fraudulent reporting of significant events, then certainty is elusive, and faith fails to manufacture truth, only belief. Religion is risky business; a lot rides on the authority we confer on others. Maher is correct on this score.

Mention must also be made that atheists are not the only ones to speak approvingly about evolution. Some evangelicals dismiss it while others embrace it. Allow me to cite an instance of the latter. Self-proclaimed evolutionary evangelist Michael Dowd experienced a revolution in thinking. Much like we might adopt a perspective that for us had gone underappreciated far too long, and we may be kicking ourselves for being so obtuse, Dowd came to terms with what some other evangelicals reject as their mortal enemy, namely standard Darwinian thought. As outlined in his volume entitled *Thank God for evolutionary theory*, Dowd finally recognized that accepting evolution does not require abandoning religion. In his view, religion can be enhanced through biology, implying that evolution can be received from God's hand as a good gift. Kudos. May his gene pool flourish.

Dowd's modification of perception generated an understanding of evolution as an expression of God's ongoing creativity. One needs only a cursory look at the biosphere, the world of life, he says, to notice the extravagance of God's providence in the shapes life takes. Dowd sees God's blessings in the richness of life and the fullness of its manifestations. Where life has a chance to thrive, there one will find some form of it. If there is a niche to be made, an organism, sometimes remarkably, will find a way to fill it.

On the surface, Dowd's view sounds like an amicable accommodation to scientific advancement. Yet his positive approach might not only be hasty but even Polyannaesque, where evolution and the divinity behind it, and bestowing blessings upon it, can do no wrong. He even refers to the evolutionary process as glorious and the knowledge of it as a saving grace (Dowd, 36, 15). He does not overlook its negative aspects, he just does not highlight them sufficiently.

In actuality, natural selection is messy and comes with much waste and suffering. The answer that evolution has for the extravagance of life is the perishing of weaker members and ultimately the extinction of those species for which there is no longer room. Extinction is the fate of all organisms; not even our species will escape it. What Dowd describes as glorious also has its brutal side. In times of food scarcity, for example, pelican chicks have been known to kill their weakest (often the youngest) sibs in order to reduce competition. As the lyrics of Ian Anderson, lead singer and flautist of the British rock band Jethro Tull, so aptly portrays this deplorable situation, which some people emphasize as the blessed work of the good Creator, "He who made kittens put snakes in the grass/ He's a lover of life, but a player of pawns,..." ("Bungle in the Jungle," from the album *Warchild*). The diversity of life can be exquisite, but so too can pain.

Since pelicans are not moral creatures, we cannot judge their actions as cruel or heartless. Dowd would need to determine, however, if the divinity has exhausted all the options, leaving the current state of affairs as acceptable. Can the deity take pleasure and pride in the blood that gets spilled when certain organisms must satisfy their complex protein needs? If this is grace, then we may require an injection of mercy to offset it. Unless of course God also thinks that the status quo is messed up and seeks to renew it. As it stands, evolution is ambiguous in that it can be assessed in positive or negative terms. Some see it as resplendent, others as at best indifferent and at worst vicious. The beautiful in one circumstance occurs at the expense of the tragic elsewhere. Is it not warranted to wonder if the cost is too high? Would Dowd be prompted to revisit his estimation if *Homo sapiens* were the ones about to go extinct? At the very least, he would have a smaller audience to do the reckoning.

The bottom line is that Maher has accurately identified his target, though it is a readily apparent one. The problem is that not every religionist fits the mold. The impression that his other target, namely his audience, likely receives is that Maher's vitriol can be applied to all religionists in all times and places. Yet there is a broad spectrum of religious devotion, not all of which Maher has taken into account. As Chris Hedges makes plain, "The new atheists, like all fundamentalists, flee from complexity. They can cope with religion in its most primitive and abusive form. They are helpless when confronted by a faith that challenges their caricatures" (Hedges, 34) and facile, glib characterizations. Some religionists are thoughtful and their faith is the product of rationally exercising their minds, not retreating from them. These thoughtful religionists might not all reach the same conclusions, but their endeavor yields tangible fruit and also pays spiritual dividends. Maher and others suspiciously assume that participation in religion spells disaster for mankind, but in many instances a religious dimension and motivation leaves the world better off than without it. As with most things, it becomes a matter of perspective, and perspectives can become jaundiced. Our perceptions are rooted in our presuppositions and expectations about how the world is ordered, none of which objectively capture the actual state of affairs because they stem from our limited vision. We may believe that we see clearly, but we do not. Maher's analysis is an example of yet another such distortion.

Yet here we are attempting to correct this reduced visibility. Where does this leave us and our natures?

Conclusion and Personal Reflections (or Reflections on the Personal)

SCIENCE IS EXPLOSIVE, literally and metaphorically. Each discipline has its own version of explosions. These range from the "hard sciences" of physics and chemistry to the "softer sciences" of biology, anthropology, psychology, sociology and so forth. The designations of firmness stem from the extent to which each discipline is dependent upon or employs mathematics, though math itself is not strictly a science in that it is non-empirical. Math conducts no experiments, has no laboratory and is a purely rational undertaking. Yet its use makes physics the hardest of the sciences. Most other sciences suffer from "physics envy." The harder the discipline, the more credibility and respectability it garners, at least this is how physicists themselves might put it.

Back to our point. Explosions are not difficult to locate in physics, especially if it all began with the biggest of all bangs. Contrary to Newton, since there was no pre-existing receptacle or container for big bang material to be jettisoned into, space was created as the particles were ejected. "Places" were created as matter was "displaced." But explosions did not stop there. The big bang accounts for only the first three elements in the periodic table, namely hydrogen, helium and lithium. It was not hot enough for long enough to manufacture any more. Interiors of large stars, stellar furnaces, were required for those elements up to and including iron. Nuclear fusion reactions, generating explosions, arrange for these higher elements. Then supernova explosions account for the remainder up to uranium. The rest of the elements in the periodic table do not occur naturally but are synthesized in the lab. With this the stage is set for chemistry, no stranger to laboratory explosions.

Continuing to move forward in time comes planet earth and the biological or life sciences. There is an explosion of sorts that occurred about 525 mya and is labeled the Cambrian explosion. In it, all the phyla (the basic body plans) that were ever attempted in multicellular organisms transpired. Most of these were lost. The few successful ones that survived yielded the ten to twenty million species we have in the present biosphere. The Cambrian explosion was the time of greatest disparity, meaning types of structural forms, which has since decreased. Conversely, since that time the world has seen an ever-increasing diversity, that is the sheer number of species. This truly was a figurative explosion felt around the world. Some are tempted to apply this to the Genesis account, which seems to be speaking the language of diversity, where God creates many species which persist until the present. Yet the Cambrian explosion could not be in view here since most of those forms became extinct long ago. Nice imagery, though.

The next explosion comes from anthropology. Anatomically modern humans have been around for about 100 ky, while behaviorally modern humans enjoy a history half that length of time. The origins of language and religion stem from the beginning of that same behaviorally modern time period. Are we justified in thinking that this is a period when humans became self-aware, self-conscious, existentially cognizant of mortality, developed the concept of a soul or otherwise had a great spiritual awakening? Was it then that their recognition exploded and flowered, manifesting a preoccupation with the transcendent, thereby allowing them to earn their supramundane spurs, stripes or wings? This again differs from the Genesis account in that Adam and Eve had no forebears. Rather, the natural inclination of humans had a

Conclusion and Personal Reflections 171

long evolutionary history and then at some stage we were able to reflect on it. There was no build-up, however, for the first biblical humans, since they came ready-made with a nature that we are still attempting to unravel. Appearing on the scene, as they did, without a history to draw upon, like Locke's *tabula rasa* (blank slate), they must have had a lot of questions. I can envision that after a long day tending to the Garden, they might have asked God, "Did you need to make lower backs so susceptible to pain?" Or did that arise only after the Fall and curse?

The final notable explosion for our purposes here is a historical-cultural one. In the period extending from about 800 to 500 BCE, there came on stage another explosion, this time of thought, specifically in terms of religious and philosophical thinking. Several religious philosophies had their origin here. This is a time referred to by Karl Jaspers as the Axial Period, where significant religious figures such as the prophet Isaiah in Israel, the Buddha in India, Zoroaster in Persia, Lao-tzu in China and Protagoras in Greece, as well as the texts they inspired (together with the Upanishads), arose and began to influence their world. The amazing coincidence, call it a different type of ID, is that these movements surfaced when the individuals in question had virtually no contact with each other across cultural borders. Greece did not really engage as yet with the Ancient Near East or Persia (until Alexander the Great's foreign policy modified the situation), nor they with the Indian sub-continent, nor India with China. All of that came later. What then was it about this time that made it such a fertile era for the human religious and philosophical imagination? The Axial Period saw the blossoming of human genius in textual form. Humans here witnessed a literary explosion.

I am reminded of the monolith in the film *2001: A Space Odyssey* that assisted proto-humans to become tool-makers—a good intention, despite these tools also being used as weapons of war. Nevertheless, my question is could God be interpreted as operating as a catalyst for the explosions outlined above? Perhaps this is how God coaxes structures and behaviors, form and function, out of the world, thereby asking, even challenging, it to step out and branch out (a not unwarranted metaphor given the sprawling bush description of species life on earth) with courage into uncharted territory. If so, God would be not only a spectator but also a participant in evolutionary and cultural history.

Recall, as I mentioned earlier, being human is insufficient for guaranteeing both image-bearing status and longevity, for there were at one time as many as three human species cohabiting on the planet. Only we have survived through to the time this volume goes to print. Why is that? To reiterate, either not all humans were image-bearers or, if they were, image-bearing is not enough to warrant God's special favor. Jewish tradition has a healthy self-image of being God's chosen people, somewhat arbitrary though the choice may have been. We cannot be certain what it was about the Jews that made them the object of God's affection. Evidently, there was something attractive about their having been few in number (Deuteronomy 7:7–8), suggesting perhaps that God likes to cheer for the underdog. The same can now be said not about a race of people but a species. If all humans automatically bear God's image, if this is part of what it means to be human, then as only one human species survived, was this selection arbitrary as well? What was it about us that held God's attention (as per Psalm 8:4-6)? If our analysis is correct, then it seems there was a chosen species before there was a chosen people within it.

Is character built or revealed (or even concealed)?

It is easy to become pessimistic about human nature. We are the reason, after all, that laws are required. We either do not know how to relate to others or are unwilling to do so; hence codes of conduct and criminal codes are drafted. Prisons are full to overflowing and courts of law are overburdened with cases; all of which testify to the fact that humans do not place a premium on conflict resolution. Not only do we often fail to love our neighbors, we sometimes wish that they would just evaporate. Self-interest appears to be the culprit.

Using biblical applications, we are the reason that the Ten Commandments had to be written (Exodus 20). Specifically, we have become devoted to God-substitutes, ranging from an overemphasis on our work to the same for our resources. Our families sometimes receive insufficient focus. We take what is not ours, even something as simple as not putting in the full amount of time at our place of employment. We do not always speak the truth, and we are envious of others. Greed and competition with our peers can be our motivation for acquisitions.

We are the reason that the *lex talionis* was written ("An eye for an eye and a tooth for a tooth…" (Exodus 21:24; Deuteronomy 19:21)). We have been known to act vindictively and to seek retribution and vengeance of the type that is disproportionate to the perceived offense.

We are the reason that the third chapter of Ecclesiastes was written. The list of all activities for which there is a season includes hate, war and killing, about which we have become proficient agents.

We are also the reason that the Sermon on the Mount (Matthew 5–7) and the Plain (Luke 6) were written. The default drive for all of us, to say the least, is not always geared toward thirsting after righteousness, showing mercy or being peacemakers. We tend to be more prideful, boastful, assertive and aggressive.

Humans are capable of both the height of compassion and charity as well as the depth of grievous evil and egregious injustice, along with everything in between. We have been known to sacrifice self for the sake of others, sometimes perfect strangers and even enemies (as in dying for one's country, not all the inhabitants of which are our friends, in wartime), in addition to sacrificing others for the sake of self (as in eugenics programs). We are ambiguous this way.

But the prospects are not all bad, for God has injected a measure of the spirit so that we might be renewed. God has invested too much in the world to scrap or abandon it; instead, God would rather "fix it than nix it."

Religion and science: Any hope for a truce?

Why, we might also ask, has so much strenuous effort on the part of some been put into the eradication of religion? I suspect that humans shy away from responsibility, as portrayed in the passing of the buck in the mythical Garden of Eden by the three particulars of Adam, Eve and the serpent. We would rather not be accountable to some higher power and answer for our deeds. The farthest we ever hope to be the recipients of just desserts is on Santa's list, trusting that no one actually makes it onto the naughty side of his ledger. Maybe we have authority issues and would prefer being beholden to none. Science figures in to these calculations, for if everything, say, is material, then we can be free of a divinity once and for all. At least that is the (non-material) mentality. Thus can there be a suitable relation between religion and science, or is it undermined from the outset? The question is worth exploring and the attempt has been made only on a few fronts. John Haught's typology is one and Ian Barbour's another; we will probe the latter's offering.

Barbour contends that there are at least four ways in which religion and science can interact; the alternatives are conflict, independence, dialogue and integration. Conflict is the form which most readily manifests itself in the contemporary mindset. With knee-jerk reflex speed, many suspect that conflict is the only fathomable reaction that the two can have toward one another. The temperature will rise whenever they come in contact. How else could it be? Robed clerics versus lab-coated scientists (or are the latter robed as well?). This perspective can be traced back to the champions of intellectual freedom who refused to be shackled by spurious creedal assumptions. Galileo and Darwin are the obvious historical examples, but these are not so clear cut and others are not so easy to find. The church authorities at the time of Galileo agreed, though not openly, that views were moving in his direction, only the time was not yet ripe for broadcasting them. And there were creationists at the time of Darwin both inside and outside the scientific camp, together with evolutionists inside as well as outside theological circles. We have the media and amateur historians to thank for the hype over conflictual relations between religion and science. They are more interested in stoking the flames than in authentic and nuanced investigative reporting, for historical scholarship does not support their thesis.

Have no fear, though, for there are other options. Independence reveals that religion and science are totally dissimilar, hence there is not even an opportunity for conflict. There is nothing about the two which would be contested, for there would need to be some point of contact first, and there is none. In the conflict type, the two are rivals setting up a frontal attack on each other; in independence, military metaphors are inappropriate. Instead, they are neighbors with a high fence in between. Each side is compartmentalized and keeps to itself and does not involve itself in the other's business. The scientist can even attend religious services on weekends provided that his or her rationality gearbox is in neutral. And the religionist might have some important ethical but not scientific things to say.

The barriers begin to erode, however, upon reaching the dialogue position. Here science and religion can interact in a civil manner. Whereas in conflict the two sides have opposing answers to ultimate questions, here is more of a humble recognition that there are as yet no confirmed ultimate answers to be had, so neither side needs to feel either superior to or threatened by the other. They can meet together to deliberate on ultimate questions where both can contribute, for each is cognizant of its limitations. Finally, integration takes it a step further, where disciplinary borders may not remain intact. Neither might remain the same after an encounter, for science can be undertaken religiously and religion scientifically. Perspectives on both God and the world can be transformed; an evolutionary world may yield an evolutionary image of divinity (Barbour (2000)).

The positions can be charted as per Figure 2, and the order will require adjustment. As one moves from left to right there is increasing contact, even overlap, between religion and science. At the far left are two circles separated by a line and this arrangement represents the independence position. Religion and science here are completely separated from each other. The next arrangement to the right is the conflict position where there is a point of contact between the two, which is highly contested and where friction is produced. This is where the battle lines are drawn and where the two can contentedly resume hostilities. Next comes dialogue where there is an amicable overlap and the two mutually benefit each other and even cooperate. Lastly is the integration viewpoint where there is more extensive overlap but of the type where the two disciplines can become changed by the influence of the other.

Figure 2. Ways to Relate Religion and Science

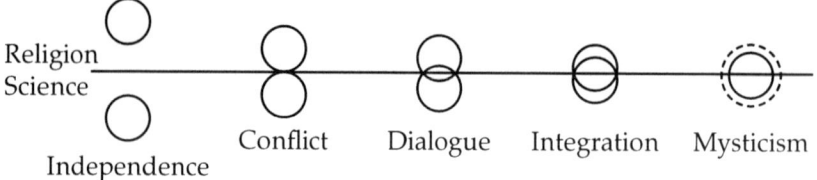

Perhaps beyond integration lies a fifth position where religion and science are superimposed. This could be termed a mystical viewpoint where the distinction between the two is blurred. They remain distinct in the other positions, though not necessarily intact in integration. Mysticism, on the contrary, might press for a unity where possible.

It could be stated that at least three opposites can be detected in this fourfold typology. Independence can be understood as opposite to conflict in terms of tension or territoriality; independence having none, conflict having much. Conflict can be seen as opposite to integration with respect to enhancement of positive relations between religion and science; conflict having none, integration having much. And finally, independence can be perceived as opposite to integration with regards to interaction or agreement; independence having none, integration having much.

Two additional comments: in conflict there are answers to ultimate questions, but only one can be right, for to accept one is to abandon the other; and whereas in conflict there are firm but competing answers, in dialogue there are only speculative, conjectural and potential ones.

Thus as consumers of investigative reporting we need not be consumed by the portrayal of religion and science dictated by the media; relations are more numerous, hopeful and show more promise than the meager helping from what the media offers.

One criticism of the science and religion debate, however, is that if there is any influence of one upon the other, then it is to be found in the science to religion direction. That is, science can have an effect on religion but religion does not really affect science, so the impact is, for all intents and purposes, unidirectional. Granted that the influence is easier to identify in the science to religion direction, for I myself accede to advancements in archaeology, as an example, as having a shaping influence on the historical aspects of Christian thought, suggesting, as it does, that the biblical accounts of the Exodus and conquest of Canaan are largely mythical in nature. This, of course, if actually the case, does not undermine faith, but makes it more informed and helps us also to appreciate the literary capabilities of the authors. Yet to claim that religion remains silent and without contribution in the science direction is hasty.

As in my first volume with Blue Dolphin, I examined several metaphysical/religious schemes which impinge upon scientific methodology, and it seems appropriate to list some of them here. Pierre Teilhard de Chardin attempted to impress upon us that there are two types of energy operative in living organisms: a tangential energy of the kind that conventional science investigates, where external relations, or the outward effect subjects have on other subjects, are in view; and another that science has failed to recognize, namely a radial one, where the interiority of an organism drives its own evolutionary process onward, upward and outward from its center. While not readily apparent, this latter form of energy calls for our attention, for science, Teilhard argued, is incomplete without it, and it was his religious views which informed him of it. And if we substitute metaphysics for religion, the next three views also apply.

James Lovelock, in another proposal, urges that our planet itself, and not only the life forms on it, should be accorded organismic status. The earth bears the marks of a living entity and science should be modified so as to include "geo-physiology" and "planetary medicine" in its research programs. Next, Rupert Sheldrake prompts science to recognize an additional category of inheritance that organisms encounter, namely morphogenetic fields, that assist in shaping the morphology, or form (as well as behavior), of an organism through the formative causation of morphic resonance—the shaping influence of previous forms onto present ones through the impact of repetition. If repeated sufficiently, forms and behaviors become habits (taking seriously the term habit-forming). Finally, David Bohm impressed upon us the view that to regard our own order of existence, the explicate order which science examines, as the only one to occur is short-sighted, so he introduced another one—the implicate order. Forms are folded into the latter and unfolded in the former, using a holographic image as a model.

The upshot of these proposals is that science would be impoverished if they were not taken into consideration, not least of which is the explanatory power and insights which they afford. Admittedly, the existence of what these researchers propose is not obvious, which is why they seek experiments which can corroborate them. By the same token, science finds itself in a similar situation, hypothesizing as it does a vast reservoir of dark energy as the major component of the universe, though undetectable in principle. Another example is superstring theory, which offers not observation but mathematics (a rational and not an empirical tool) in support of its claim. The amount of faith invested in the unseen is truly inspiring and demonstrates that science has its religious moments as well, and that faith can drive the scientific enterprise.

The good, the bad and the ambiguous

Are we now in a position to decipher whether the makeup of humans is basically on the positive or negative side of the ledger? Allow me to approach the question in this way. The same God who declared that God is not interested in animal sacrifices (1 Samuel 15:22; Psalm 40:6; 51:16; Proverbs 21:3; Hosea 6:6), as though they are held to be a method of appeasing the divinity, announces that God's primary objective concerning humans is to renew their heart. This suggests that renewal is warranted and that God cares little about the formalities of ritual observance, what might be termed religiosity. So how can humans be understood? They are like a garden containing weeds, the good is mixed in with the bad. There are moments when either seems to dominate, implying that we cannot always determine how we tip the scales until the final reckoning. In the meantime, we are called upon to team up with God to extract the weeds, realizing that this is a never-ending task.

There can never be a thoroughgoing house- or yard-cleaning, for weeds have a way of infiltrating even a well looked-after plot of ground. Once again, we are ambiguous this way, and it depends on who is doing the reporting—us or someone else.

It has been claimed, curiously in my estimation, that everything has its purpose. If so, then why do the scriptures declare that weeds are good only for being plucked up and thrown into the fire (Matthew 13:40)? I feel the same way every spring about dandelions on our lawn. They offer a scant natural food source for the creatures that I am aware of, outside of the otherwise indiscriminate bees that are attracted to the radiance of their amber glory and elect to drop in for a visit. And there are other plants from which to make wine. If they were not to exist, no organism would be much the worse for it. Hence are dandelions bad or ambiguous? My lower back at least maintains a strong conviction about the former.

Thankfully, we are not left without assistance in the endeavor to transform our lives. God's spirit makes it a team effort. And in response to Kant's assertion that ought implies can, we might say "in part." Left to ourselves, we may not undertake the task of renewal as completely as needed. We must do our part, yet under our own steam we remain a necessary but insufficient condition.

Armed with the foregoing, consider the following. Continuing to recall themes from a previous discussion, arguing for a biological basis of memory has some difficulties, for it could only be traced to something structural or chemical. If the former, then memory would be dependent on a specific array of neuroanatomical structures, such as a constellation of neurons. But as we affirmed, they can become altered through the adoption of different decisions. Under these conditions, could memory change if the structure changes? Recall we are asking if memory is based on structure, hence would choices then make memory fluid? This does not seem to occur, so we might need to search elsewhere. Or if the latter, should memory be encoded and stored in neurophysiology and its biochemistry, well a significant amount of our makeup becomes recycled periodically. Some aspects and processes undergo such change more so than others: epidermal layer and intestinal lining rapidly; myelin sheaths surrounding neuronal axons slowly. Can memories then be similarly affected? They could be if they are closely connected to body chemistry. These are some of the potential pitfalls of a purely substance view of memory.

Having covered this ground, we can ask about neuroplasticity as applied to lifestyles. For religious purposes, specifically in terms of the Christian West, there can develop a redirection of life with deleterious choices becoming salutary ones, which, employing the language of Sheldrake, if firmly entrenched through repetition become wholesome habits. The more consistently these thought and behavioral patterns are selected, the more anatomical (dendritic) pathways are channeled to accommodate them. Different signals forge different pathways and God seeks this type of positive transformation, regeneration or reorientation in individual lives. We have a plastic brain that is always in the formative stages. We never grow too old to construct a different us.

New and improved?

God is interested in renewed lives; now is there historical evidence to be had of such an outcome for the human condition? More so, is it possible to graph human moral history, by any or all accounts, so as to determine if it is or has been in a state of improvement or decline? Here are a few attempts, at least by the standards of how some might describe it (see Figure 3).

In the first instance (1) is what could be referred to as the classical literalist interpretation of the Judeo-Christian scriptures. Here humans together with the rest of creation are not perfect but very good, as per the end of Genesis 1. This (a) lasted for a while until such time as the serpent got the better of the first human pair, causing sin and death to enter the world, and provoking God to place the creation under a curse (b). Humans were then on a steady decline until God had had enough and sent a deluge to wipe out all but a remnant, who would then repopulate with more godly offspring (c). Events proceeded at an even keel through this stage, with evil proportionate to the size of the population. That is, the moral fiber neither increased nor decreased over long periods,

Figure 3. Graphing the Human Condition

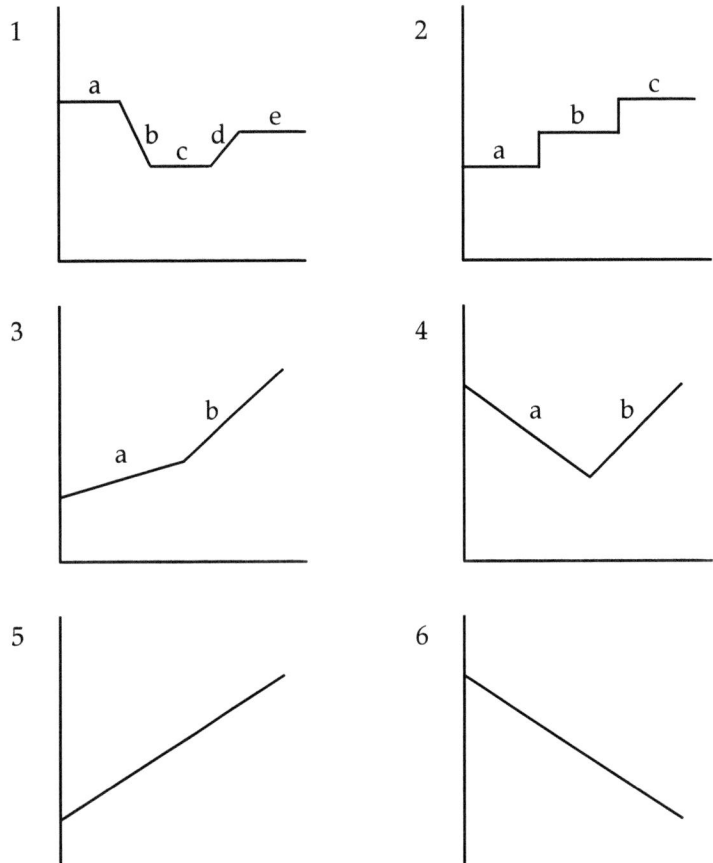

though occasionally God had to send prophets to get the people back on track. Then in the fullness of time Jesus enters the scene, inaugurating God's kingdom (rule or government), and paving the way for redemption and reconciliation to God (d). At best, moral fiber later plateaus (e) but does not reach the initial Garden level; for that, Jesus will need to return in order to fully and finally set up God's kingdom on earth. Resurrected humans (let's call this stage (f)) would then be on a higher moral plane.

Problems with this approach include the following. First, there is no anthropological evidence for any Golden Age of humans, only a steady and gradual natural selection process which ultimately produced humanity, and with much blood on the ground along the way. Proponents of (1) therefore reject the evolutionary picture. Second, if the work of Jesus means that the end is better than the beginning, then the death of Jesus becomes a good and necessary thing, an integral step to get from (a) to (f). The question then arises, why not start off with (f) so that neither the curse nor Jesus' death ever needs to occur? Otherwise our sinfulness becomes a requirement, without which (f) could not be achieved. So the sooner the first humans sinned, the sooner the redemption ball could get rolling. I suppose we should heartily thank the serpent for this; a concession the literalists would not be prepared to make.

Another possibility (2) is that the amount of evil and sin remains constant (*per capita*) in the post-diluvian world (a) until such time as Jesus becomes a catalyst for a change in the proceedings (b). Humans continue on a wayward path, but the situation improves once God draws history to a close (c). The change is abrupt at each stage.

Problems with this rendition: it assumes that human morality matches human consciousness. Moral awareness translates into moral response, with positive and negative in roughly equal proportion. At least this view unwittingly agrees that we are actually less warlike than our romanticized evolutionary ancestors (Ehrlich, Ward). Some anthropologists would wish to place (a) on a downslope; others who accede that we have grown less bellicose would push for an upward trend. Literalists, as not part of the conversation, might not recognize a distinction. However, what would effectively disconfirm (1), there is no clear indication that the moral fiber of humans is in fact on the rise since Jesus' ministry (1(d)). History does not bear this scenario out, thereby making (2) a more accurate picture.

In (3), humans are always improving (a), and even more so since the influence of Jesus (b). Yet once again, where would the conclusive evidence be for the incline in either (a) or (b)? In (4), humans were declining until Jesus came. Sadly, the evidence is still lacking. In (5), everything is on a steady incline over the long term. This approach takes evolution seriously and, according to the views of some, interprets its history accurately. But at what point, if any, would the injection of Jesus have made a significant impact, and does morality really proceed apace with evolution? Lastly, in (6), ever since humans either received externally or generated internally a moral sense, they have used it for their own ends and in blatant disregard of others, Jesus notwithstanding. A final alternative, in my view, would be a horizontal line, indicating that humans have remained much the same since their inception. There may be periodic crests and troughs as in a sine wave, depending on the circumstances, but over the long term the shape smooths out. This is the scenario that I envision.

There could also be either some as yet unimagined graph more germane to the actual state of affairs, or the history of human nature and morality simply cannot be drawn so easily. Much of the analysis depends on one's perspective, especially if the issue of human good or evil has already been settled in the minds of the perceivers. A further concern is the interpretation of moral responses and whether they reflect positively or negatively on human character. The difficulty is that we cannot even agree on which side of the ledger specific actions are to be placed. Many understand helping the downtrodden to be commendable; Ayn Rand would not concur.

Regardless of the position one takes on the foregoing, it is fair to ask if the event of Jesus' life has translated into an improved human condition and has made us reflect his character in increasing measure. His life and work might be seen as a definitive act on the part of many, for they would answer in the affirmative, but has it made the angle of incline any greater than before? If not, then was his example merely a signpost for what the end of history and its aftermath would be like? Some would affirm that he has had a lasting effect on their own lives, and this is the essence of conversion, yet was the world in general better or worse or the same prior to his arrival, and how could we tell? In any case, one would hope at least that his and their efforts produce fruit both now and later.

In my estimation, humans are progressively inventing ways to be devious, for we "have devised many schemes" (Ecclesiastes

7:29). We do not always acquit ourselves well as a species and this does not restore my faith in humanity. Every time there is a new technology, for instance, some people will find ways to exploit it for harm. No doubt the ancients operated similarly; we just have more avenues for it. Technology is certainly used for much good, but which outweighs which? Like a black stain on a white surface, the white far outweighs the stain, but what do the eyes focus on and what do we remember? The bottom line for me is that the good abounds more than evil, but the evil is louder.

On revelation—God's and mine

It is a rare occasion when the world of experience, rather than science outright, begins if not to corroborate then at minimum to speak to theological issues. I refer to circumstances that would be informative if not also telling about the likelihood of God showing God's hand. Perhaps by the end of the current century the population of the planet, should it continue on its present course, will according to the Malthusian principle reach a point of unsustainability. There will not be enough room to produce food for all who need it, unless of course we learn to farm the oceans more adequately. If and when the human population comes perilously near this brink, there will be nowhere to go but a diminution of our numbers for lack of food or increased diseases or both. Similar reasoning would apply to extraterrestrial threats such as those asteroids with our biosphere's name on them. One wonders if God would leave us in such dire straits without assistance. Should these be the circumstances when God draws history to a close, then the timing would be good. If not, then God might have us fend for ourselves and allow natural selection to take its course and weed out the unfit (in a passive way, whereas culling the herd would be active and harsh).

Indeed, if the human race itself becomes threatened, could we expect God to act and even anticipate some correction of our plight? Or is this notion the fruit of presumption and self-importance? Do we have the right to expect that God would bail us out as a species particularly when the difficulty is of our own making? We have obeyed God's command to be fruitful and have multiplied to fill the earth, but there is a limit. Should we have been better managers of our fruitfulness and multiplication propensities so that our numbers would not become problematic? Or are we doing the right thing by God's lights? We may not know until the line has been crossed. If God does not show God's hand and act by then, it might be too late. Human species have gone under before.

So much for a divine revelation, now here are a few additional disclosures of my own. First, on the topic of perspective, my view is that people are sometimes under the mistaken assumption that they see clearly while everyone else suffers from myopia or some other ophthalmic disorder. Our own perceptions are believed to be the accurate ones and we insulate ourselves from potential revision by not entertaining the perspectives of others. Humans are prone to shy away from the effort required to see from another angle, for why make the attempt when your own suffices? As long as a position is challenged only from without and not within, it may become entrenched and difficult to dislodge.

There is applicability here with the Christian West and its biblical tradition, thereby launching our first of four extended treatments. We are selective literalists in that we have our favorite biblical passages which we insist on in a way that may as well be literal. For example, one group that does not consider itself to be literalist, namely process theology, takes the "in [God]" of the passage in Acts 17:28 ("For 'In [God] we live and move and have our being'") with the same degree of seriousness as a literalist might, whether or not that was the intent either of the apostle Paul or the Greek poet Epimenides of Crete who first penned the words. We build theologies around our pet passages, making them of greater importance than the rest. Martin Luther based his Christology (the doctrine of the nature of Jesus) on a disputed passage, which could read "God was in Christ reconciling the world to himself" (2 Corinthians 5:19), or alternatively, "in Christ God was reconciling the world to himself" (NRSV) or "God was reconciling the world to himself in Christ" (NIV). Each presents a different idea about the person of Jesus.

And on the theme of how the scriptures are put to use, there are those who would insist that to accept them, and indeed the Messiah to whom they point, is a liberating experience, "For [his] yoke is easy, and [his] burden is light" (Matthew 11:30), and amounts to the truth that will set us free—the freedom in Christ. This is the promise believed to accrue to the faithful if they follow his lead. The trouble is that there are strings attached to this evangelical form of Christianity, namely the orthodox view of the faith. Here the scriptures are understood to convey the message of God in a way that does not fall victim to the pitfalls of mere mortal communication. Not only the very voice but the very words of the text are held to be inspired or even inerrant and infallible.

The history of textual criticism, however, has demonstrated that the human transmission of the text is not immune from the broken telephone effect, whether or not God is involved. Plus, additional doctrines (read dogmas) are assigned for the neophyte to adhere to. This begins to sound burdensome after all. Upon reflection, the disciple might come to the realization, through the use of the mind that God bestows, that the Bible and the doctrines drawn from it are not as airtight as previously hoped for or assumed. This new mindset might prompt another move, namely the embracing of a more nuanced approach to the faith where former absolutes become relativized and a greater appreciation is gained for other viewpoints. If that can be adequately described as a freeing experience, then the liberation that was initially promised apparently comes in multiple steps, for the first one no longer delivers. Perhaps, on the one hand, freedom in Christ is not sufficiently facilitated in institutional settings, necessitating a liberation from them as well; and on the other, maybe growth and development into maturity requires additional phases, indicative of the educational process, and is not manufactured ready-made in one fell swoop. Like many things, faith might take time to age properly. Yet at each stage there are presuppositions that we commit to about who we are, what the world is like and our place in it. Essentially, we substitute one harness for another every time we take a life-sized step, hopefully in increasing order of liberation. In this case, we might become a slave to our relativistic underpinnings, preventing us from dropping anchor in some definitive intellectual harbor. But make no mistake: we can never escape from some type of fetters, for don them we must, so let us choose wisely.

Second, contrary to some traditions, we do tend to hold to the notion of a self and even its location. It takes effort to deny our self, and we betray where we think this self resides in the act of pointing to it. We use our index finger to point to a specific body part when referring to our self, and the sternum is usually that place which constitutes an "I" or a "me." By our actions we admit to self-ownership.

Third, our next lengthier piece is devoted to Darwin's legacy. Natural selection is not for the squeamish. There are no warm fuzzies to be gleaned from a purely pragmatic mechanism. If a trait works, it gets used; if not, out it goes. If you can't handle extinction, then don't bother, as natural selection is not for you. As a description of how we and every other organism got here, natural selection is entirely opportunistic. It seizes the moment by employing what it can to facilitate organisms as going concerns. If the attempt fails, then the unfortunate organisms are relegated to the dust bin of history. This is not the place to look for compassion, for natural selection does not go about its business ethically. Its prime directive is to pass genes on to the next generation, even if it means doing so by what we might consider immoral means.

Now the argument can be advanced that morality has arisen because it assists in the very program to which we have just alluded. Two points appear to be at play here: "That much of [our] behavior is brain-based is suggested by findings showing that humans suffering from [certain brain] damage…are sharply transformed from their pre-injury moral judgments. Part of the morality machinery is broken" (Tiger & McGuire, 121); and by the same token, "It is unclear that, by themselves, the neurons could evolve any ethical system" (Beauregard & O'Leary, 152). Regardless of the mechanism by which it has appeared, morality builds community and is good for the group, implying that a moral sense has survival value. But do not assume that natural selection is thereby up for some humanitarian award, for it is more akin to the business world whose main interest is in profits. Corporations that, say, produce green products may not actually have an environmental conscience at all but might seek to ride the wave of a trend that could translate into sales. Natural selection is all about business, so don't take it personally.

If you prefer a more benevolent picture, then consider a moral divinity who pursues the path of compassion and hopes that creatures like us will do the same, and who seeks the best for organisms and accomplishes this by coaxing out the most beneficial variations for them to bear. Mutations by themselves cannot be relied upon for the best outcomes, so God stacks the deck or loads the dice, ensuring that beneficial variations have more than a level playing field to work with. As in a parent-child relationship, God seeks to give offspring every advantage in order that they can make it in the world. But the process is experimental and can largely be described as trial and error. Hopefully, though, God's approach produces much less error. It could also explain how evolution has made the strides that it has in the time allotted. From rodent to Rodin in a mere 65 my, ape-like creatures to Nobel laureates in 5 my. It might be a stretch to portray this history as gradual. To my thinking, this rate bespeaks divine assistance (though not design, for the architecture of my lower back sometimes barks out a different story).

Finally, if one of the objectives for all animals is to leave robust offspring, then those organisms that feed on their young are maladaptive. Evidently, they were not consulted in the deliberations. Besides, how can we be certain that reproduction is the main motivation for most creatures? Maybe they just want to mate and have not as yet worked out the cause and effect relationship between mating and offspring. Beyond this, once progeny appear, care for them might kick in as well. Nevertheless, parental instinct is not guaranteed.

Fourth, even inveterate views can undergo a renovation, for I find myself reverting to a dualist posture. I neither devalue the material world nor denigrate the physical body as part of it as a Platonist would, but I do think that there are at minimum two classes of things in the universe, namely matter and spirit. I agree with Powell that "consciousness appears to be neither energy nor matter [as is the brain] but a field." As we have already established, "changing our thought patterns can rewire our brains, and… changing our brains' wiring will alter our thought patterns." The implication is that "Consciousness acts like a force when our thoughts change our brains' wiring," and this is akin to telekinesis (Powell, 204, 208, 210). There do seem to be at least two components to our makeup after all, the physical and the mental. I consider the idea of our being psychosomatic unities as merely wordplay, which by itself does not overcome the problem nor unravel the complexities and intricacies of the two items that are deemed to unite. Nevertheless, as is the dualist frailty, I concede that I am still mystified as to how the two can interact.

This is an important issue and deserves an extended look, marking our next lengthier item. As mentioned, dualism need not be of the Platonic variety. The apostle Paul emphasized not only a moral dualism but an ontological one as well. We can list these statements: we live in an earthly tent that will be destroyed at death (what is this "we" that lives and dies?); we are naked at death and seek to be clothed not with the mortal but with life; "while we are at home in the body we are away from the Lord" (2 Corinthians 5:1-8); and remaining in the flesh is to be removed from Christ, but to depart is to be reunited with him (Philippians 1:23-24). Recalling a previous exercise, these concepts prompt us once again to ask what the "I" is that has a body. We will revisit these themes below as they relate to the resurrection, but for the present we can go beyond these sentiments and declare that God values the creation highly and so should we, yet the transformed physicality that the new earth will enjoy is of a different type than the one currently in place. This does not devalue the current conditions but looks forward to a time when an advanced stage will replace it.

This still leaves the question as to the mind/soul and how it got here. There seem to be two possibilities. Either the soul is an emergent property and as such is a natural product, through God's assisting the evolutionary process, or, as with Alfred Russell Wallace, it is a direct divine deposit (DDD). Like a magnet is a natural product and exerts an influence through a field beyond purely physical contact, so consciousness and a mind could be produced through complex circuitry. The difference is that once magnets become demagnetized, or in our case we die, so too does their influence. This implies that there would then be no mental survival, assuming that the analogy can successfully be applied to us. If not, then we are left with the latter prospect, that what is metaphysical did not spring from the physical but from a divine source. If the natural world did indeed generate a mind that survives, then we need a different analogy; and those who object to divine intervention in the world will not favor the DDD approach.

A DDD would be similar to the Platonic view where a soul is retrieved from the realm of Forms, as one of the Forms, and is dropped into an awaiting physical receptacle. This depicts the standard dualistic posture, but it need not be this way. If God draws mind out of the natural sphere, then mind arises or emerges out of it, but whose product is it? It might come from the natural domain but is a result of divine craftsmanship and as such may constitute a third option. And if so, it might bear marks and qualities that can endure. As extra-material, it may not suffer the same fate as the material; God might see to that. Consequently, minds could arise through a natural-divine partnership and endure once the body expires.

Paul in 1 Corinthians 15 outlines that God will dispense with material bodies, which have outlived their usefulness (see John 6:63), and reunite minds with new spiritual bodies. More on this in a moment. For the present, we can tentatively conclude that dualism is not always to be interpreted negatively, for as caterpillars shed their cocoons so that butterflies can emerge, not thereby disparaging the importance of the initial stage, bodies, themselves a marvel in their own right, will be superseded by ones that will last. Bodies, along with the rest of creation, are very good but provisional; what is to come is even better. Minds were built to last; bodies were not. Yet the butterfly analogy still does not fully capture the change about to take place.

As an aside, on the topic of why it was that God took so long to coax out behaviorally modern humans from their ancestors, together, potentially, with a concurrent mind that could entertain the significance among a set of behaviors and select appropriate ones, one response is, "What's the hurry? Sometimes doing it right means taking the time to allow the ingredients to simmer."

Fifth, on the theme of design, in the developing embryo, genes are "selectively turned on and off" so that differentiation of cell types will occur, in order that cells, which all bear the same genetic complement, may be regulated into different types and functions. Powell asks what it might be that coordinates all these processes when the embryo is not working according to some blueprint for the finished product (Powell, 215). If the IDers are in error and what has just been described is not divine planning and organizational supervision, then what is it? The same of course can be said for a seed and its eventual end result of a fully grown and mature plant. Paul employs the same analogy to describe the transition from our physical to spiritual bodies. I don't much like being a dualist, though a modified one, but there I am. Now I am looking for a way not to plunge into the ID deep-end as well. Perhaps it is simply a matter of the scourge within much experimentation, that anomalies will inevitably surface to sully a perfectly tidy theory. Regardless, I admit these theoretical vicissitudes on my part.

Sixth, back to the discussion from point 4 and now our final protracted theme. For Plato, death is looked upon positively since it emancipates the soul from the passionate impediments of the body, thereby allowing it to return to the spiritual life in the realm of the Forms. For Christians and most Jews, death is perceived negatively as an enemy which (for Christians) Jesus has conquered, inaugurating both a kingdom of God's rule or government and, now as a life-giving spirit, a race of spirit-beings, and which (death) God will ultimately subdue. I mention most Jews since the ancient Sadducees did not accept the concept of resurrection. Only for Pharisees and, later, Christians would the notion of a re-embodiment have resonated. In the Judeo-Christian case, humans are pulled in three different directions, assuming for the moment a tripartite perspective on human nature. The body returns to the dust, the spirit or "breath returns to God who gave it" (Ecclesiastes 12:7) and souls or shades descend to Sheol, or the pit, to await the resurrection (Tabor, 51-59).

The transformation process of resurrection is not one of reconstituting a mortal body, for there may be nothing left to work with if the physical is fully decomposed, but the personal identity will be reclothed with an immortal spiritual body. While it remains unclear as to what the "it" is that is left naked in 2 Corinthians 5, this "I" with its former life amounts to the seed which Paul speaks of in 1 Corinthians 15 that will put on imperishability. As in the mythological Adam, all humans are of the dust and to it they shall return, for mortal death is inevitable; and in Jesus humans will rise to life with a new spiritual body. This constitutes, in my view, a more adequate understanding of the doctrine of original sin, for it depends on our inheritance: mortality and perishability from Adam, life with Jesus. Such a formulation is not in terms of our having been morally infected or marked with a stain that we cannot remove and that has been passed down to all subsequent generations. Rather, using the Adam symbolism, we have attempted to be what we are not, namely divine-like, and have consequently been barred from the tree of life, otherwise enabling us to live forever (Tabor, 61-64). There is a persistent nagging problem, though, namely Paul's reference to the perishable needing to put on imperishability. Bodies are what perish, not the "I" by the above reasoning. The "I" puts on another body; bodies do not put on other bodies, but replace them. Can we be certain then, if we insist on using these passages of his in support of this claim, that Paul has the "essential self" in mind, as that which is given new garb?

Paul's writings, which came first, emphasize a resurrection with a spiritual, though like the angelic beings, a substantial (and therefore not no) body, while the gospels, which came later, stress a physical resurrection body (perhaps in response to a perceived growing gnostic threat which devalues the body and all things material). Despite the latter "limitation," the risen Jesus seemed to enjoy a body with super-physical powers, like passing through locked doors (John 20:19, 26). In Paul's case, it is unclear as to whether the new heaven and earth of Revelation 21 will provide a corresponding earthly spirit-world for spirit-beings to inhabit. The biblical perspective is clear, however, when it comes to Sheol, for there is neither knowledge nor activity while the deceased are in the grave (Psalm 88:10-12). But I believe there to be both knowledge and activity beyond it. The new life could "involve a setting free of powers which were inhibited" previously, as well as "equally involve an inhibition of powers which were formerly freely exercised" (Broad, 430). In my appraisal, what we "take with us" into the afterlife, indicating that the saying "you can't take it with you" is not globally applicable, is our subjectivity, for this is what endures and constitutes the continuity between pre- and post-resurrection-type events and existence. In existential language, we are the sum total of the decisions we make in life, and in Christian terms this is what passes on. Oh, and by the way, humans are preoccupied with death and a potential afterlife.

Seventh and lastly, "the ability to wish or imagine that we can be better is notable. No other species aspires to be more than it is" (Gazzaniga, 389). We can rise above our natures, but this will take strenuous effort on our part. The world is messy, but we need not contribute to the messiness; we can instead work to alleviate it. And on the topic of imagination, what makes us unique includes the ability and propensity to fictionalize as well as conceptualize not only actual but also possible worlds, such as life in the future or even life after life. The outcome of the assisted evolutionary process is a creature who can entertain alternative outcomes (Stringer, 113).

I end with this disclaimer: the thing about final statements is that they tend to be provisional and views become modified with time. Informed humans are like that. Witness my reversion to a dualistic stance. Where there is uncertainty it is best not to be dogmatic, an eventually diminishing human trait I trust, but rather to make one's peace with tentativeness.

All that remains in our exploration is for me to tell you about my dreams.

Appendix: Dreamworld

WHAT I HAVE LEFT IN STORE is nothing so lofty as the "I have a dream" speech of Martin Luther King, Jr. Besides, I probably have dreams similar to many North Americans, like global peace and a Cubs World Series. So I offer the following two sleep-generated dreams as typical of what I, your human host, experience. I have few recollections of my dreams; I recall some of them upon awakening, but they quickly fade from memory. These two had staying power and I logged both before I lost them as well. I do not know what made me remember these ones from among the welter of dreams I have had over a lifetime of night journeys. Humans commonly have dreams, in that they press toward goals they set for themselves; and they do dream, whether by day or night. Here are two of the latter. Some of mine are comedies but these two are dramas.

First dream: One vivid dream I had involved "awaking" barefoot in a field in summer. I can recall thanking God that though having slept in the great outdoors I was not visited by creatures of the night, large or small. I find it interesting that the content of dreams is not looked upon as odd from within the dream but taken as a matter of course, as though sleeping in a field were a regular occurrence. We seem to readily suspend our judgment here.

I gathered all my accouterments, amazed that there was so much to carry and that I would not travel more lightly. I searched for a way back home and could not regain my mental compass. I went around a mound of sorts and noticed a hetero couple of south-central Asian extraction in an osculatory embrace. I am still unaware as to the wherefore of their cameo appearance. Psychological interlopers, I suspect.

I went on to search for a path leading out of the field and toward civilization, and found a gravel road. I walked along and came to a more modern looking library. The contrast did not strike me at the time, so I did not inquire about it. Too late now. As an aside, my dreams usually conform to physical laws in that, while I have been known to defy gravity in them as well as breathe under water, my dreamworld bodily movements have consequences if I am not careful. I am not more clumsy in my dreams, but swinging my arms around will on occasion knock something over that is outside of my field of vision, at least the direction in which I was then currently facing. I need to exercise caution even in my dreams!

I entered the building and it dawned on me that it was an academic library, serving the institution known as Ritson College. I am aware of no such campus, but if the name were to be spelled backwards, it becomes a derivative of the Latin word "noster" meaning "our." The significance still escapes me. I noticed a woman to my right who seemed older than the average student. She resembled a woman from another dream of mine. Eager to situate our educational setting in terms of time and place, I searched for a newspaper that might reveal it. I failed to locate one, so I turned to the woman and was about to ask her. Before I could get the words out, she anticipated my puzzlement and responded with, "Boston area, 1950s."

I had further questions. We went outside and walked along the gravel road together, whereupon I realized that I was still barefoot. I saw a pair of shoes in front of me and commanded that they be placed on my feet. They complied. We continued walking up a hill toward the center of campus. I don't recall obtaining any more answers.

Second dream (from another occasion): This one is longer than the first, but both probably provide fodder for psychoanalysts. I have entitled it "The Restaurant." I found myself one night, three hours prior to my scheduled time for arising, wide awake and thought, "Well now what?" I drifted in and out of this-worldly consciousness thereafter:

My wife and I relocated to a city where we had once already resided as a young married couple. The first time around we lived in a three-story walk-up apartment complex. Same thing here. We were in the midst of unpacking and cleaning, aided by another couple whose presence there seemed out of place. We knew them from a different city far away and why they should turn up here is inexplicable. Let's call them Bob and Carol. They were helping to reorient us to our new though familiar surroundings, informing us as to how the place had changed. They would not be the best ones to judge of course, never having lived there.

We were all becoming weary of the tasks at hand and recognized that we were getting in each other's way. My wife, Alice (and no my name in the dream was not Ted), and I found on several occasions that we were working at cross purposes. We came to the point where we simply had had enough. Bob and Carol were about to politely excuse themselves and depart for home when we suggested that we take them out for an evening meal. They accepted.

We asked if there was still that Mexican place in town and they replied that it had undergone several changes of hands as well as locations and that we would require directions in order to find it. They wrote them down on a piece of paper and handed it to Alice, for we had arranged for her to go on ahead of us in our car while I would organize the apartment a little more and catch a ride with our friends in their vehicle. This we did.

The restaurant was in a part of town I would not have anticipated. I spotted our car, reassured that Alice had arrived safely. The three of us walked around the building where there was patio seating, though only of the kind one would find in a fast-food establishment—not intended for comfort. There were two two-seaters available and Bob and Carol sat at different ones, asking me to join them. I figured that dormitory style was the most appropriate arrangement, making the seat opposite Bob's the obvious choice. "Why don't we eat out here?" he asked. I was inclined to pass on the idea, since the wind was picking up, we being exposed to the brunt of it, and I did not want to be a study in ballistics and watch our meals become projectiles. I wanted neither to wear my food nor leave Alice alone any longer.

"Where is she anyway?" I inquired. Bob looked around the corner to his right, seated as he was at the wall's edge, and declared "She's over there, I can see her from here." "Let's join her," I urged. But Bob and Carol were disinclined to do so. They wanted to walk around some more. I joined them, reluctantly; they were our guests after all. I did not expect hosting this event to cause such difficulty.

We walked to the corner when Bob and Carol's demeanor took a decidedly unwell turn. They began to look peaked and drew out from their satchels a snack which they kept for emergencies. It was of the junk food variety. "Why bother with a snack?" I exclaimed. "We are right here at the restaurant!" They were adamant, but we eventually made our way to the front door.

As we approached, I turned to them and both had aged exceedingly (see what junk food will do!). Both had become grey, as well as he taller and thinner, and she shorter and plumper. Both were barely recognizable. They were in good spirits but mentioned that they were no longer hungry and just wanted to return home. I said that I would pass along their regrets to Alice, perturbed that the junk food was the likely culprit of their sudden lack of appetite and that their reticence to defer satiety sufficiently foiled our plans to dine as a foursome.

I entered the restaurant and discovered that Alice had also changed in appearance, though not in age. Her face was longer and hair blonder. She lamented that the wait was growing tiresome, and felt similar to anyone who prepares an intimate meal only to find that the expected guest does not arrive. A terrible feeling, I thought, for that host does not know how to respond. Has there been an incident with the guest caught in it? Or is the guest simply negligent, in which case a reaction of anger in place of worry is understandable. Not knowing is the worst of it. I apologized and assured her that I would return in a moment, but first wanted to retrieve something from our car. I obtained it and returned to find an empty table.

I wondered where she could have gone and moved to a different table in the center of the adjoining room. I looked around embarrassingly and figured that I had better order something. I chose an appetizer and water and munched away on it, my alarm preventing me from enjoying the experience. I decided to return to the previous table and found Alice there in her normal appearance and engaged in her usual pastime, reading a book. She reported that she had been waiting there all along. I remarked that she had disappeared, at least for a moment.

She motioned for me to return to our first table, so I went back to the other room to collect the remainder of the delicacies and noticed that this part of the restaurant had emptied. I returned to our table and noticed Alice, still there, reclining on her seat as though about to take a nap. Some rambunctious youths were nearby preventing her from resting comfortably, so I went back to the other table to retrieve the water, in case I needed to douse the fervency with which they were celebrating. Upon my return I saw that the youths were gone but that the place was brimming with life. But Alice was nowhere to be found.

I turned to my right and saw a middle-aged woman nod to me, at which point Alice rematerialized. She had a rounder face and once again became the brunette from her younger days that she had long since bid farewell to. She mentioned that the woman was a friend of her family and that she calls Alice by her mother's name, Ellen, something that no one else does. So Alice had to go over and renew acquaintances. Following the passing of her mother, to be referred to as Ellen sparked her interest.

After all the transformations, I was just glad to have her sitting opposite me again. I went around the table and gave her a lengthy hug. I could feel strong emotions welling up inside me. After our embrace, she went back to her regular appearance. She suggested we leave, no longer feeling the need for anything that the restaurant had left to offer. When we came outside, ours was the only car in the lot and the building had become dilapidated. There were not even any lights around to tell us the name of the place on the sign. We used our mini-flashlights on our key chains to assist us, but to no avail. The sign was completely rusted away. No one had been here for years. I recommended that we call the place "Second Chance." She nodded approvingly.

We got into our car and began to drive. We were in luck; there were only green lights all the way home.

Bibliography

Appiah, Anthony. "But would that still be me?" In Bowie, G. Lee, Meredith W. Michaels, and Robert C. Solomon, eds. 1996. *Twenty questions: An introduction to philosophy.* 3rd edn. Toronto: Harcourt Brace & Co.: 372–76

Armstrong, Karen. 2009. *The case for God: What religion really means.* London: The Bodley Head.

Banwell, B.O. "Heart." In Douglas, J.D. et al., eds. 1982. *New Bible dictionary.* 2nd edn. Wheaton, IL: Tyndale House: 465.

Barbour, Ian G. 2000. *When science meets religion: Enemies, strangers, or partners?* New York: HarperSanFrancisco.

Barrow, John D. 2002. *The constants of nature: From alpha to omega—the numbers that encode the deepest secrets of the universe.* New York: Pantheon.

Beauregard, Mario, and Denyse O'Leary. 2007. *The spiritual brain: A neuroscientist's case for the existence of the soul.* New York: HarperOne.

Behe, Michael J. 2007. *The edge of evolution: The search for the limits of Darwinism.* Toronto: Free Press.

Borg, Marcus J. 2002 (c.2001). *Reading the Bible again for the first time: Taking the Bible seriously but not literally.* New York: HarperOne.

Bowie, G. Lee, Meredith W. Michaels, and Robert C. Solomon, eds. 1996. *Twenty questions: An introduction to philosophy.* 3rd edn. Toronto: Harcourt Brace & Co.

Broad, C.D. 1971 (c.1962). *Lectures on psychical research.* New York: Humanities Press.

Brockington, L.H. 1961. *A critical introduction to the Apocrypha.* London: Gerald Duckworth & Co.

Cameron, W.J. "Soul." In Douglas, J.D., et al., eds. 1982. *New Bible dictionary.* 2nd edn. Wheaton, IL: Tyndale House: 1135.

Clayton, Philip. "Toward a theory of divine action that has traction." In Russell, Robert John, Nancey Murphy, and William R. Stoeger, S.J., eds. 2008. *Scientific perspectives on divine action: Twenty years of challenge and progress.* Berkeley, CA: The Center for Theology and the Natural Sciences: 85–110.

Crossan, John Dominic. 1996 (c.1995). *Who killed Jesus?: Exposing the roots of anti-Semitism in the gospel story of the death of Jesus.* New York: HarperSanFrancisco.

———, and Jonathan L. Reed. 2001. *Excavating Jesus: Beneath the stones, behind the texts.* New York: HarperSanFrancisco.

Deep Purple. 1973. *Who do we think we are!* Warner Bros. LP.

Descartes, Rene. "Mind as distinct from body." In Bowie, G. Lee, Meredith W. Michaels, and Robert C. Solomon, eds. 1996. *Twenty questions: An introduction to philosophy.* 3rd edn. Toronto: Harcourt Brace & Co.: 187–91.

Douglas, J.D., et al., eds. 1982. *New Bible dictionary.* 2nd edn. Wheaton, IL: Tyndale House.

Ducasse, Curt John. 1961. *A critical examination of the belief in life after death.* Springfield, IL: Charles C. Thomas.

———. "Are there rational grounds to believe in life after death?" In Frederick E. Mosedale, ed. 1979. *Philosophy and science: The wide range of interaction.* Englewood Cliffs, NJ: Prentice Hall: 86–94.

Edwards, Paul. "Some theories about the mind." In Frederick E. Mosedale, ed. 1979. *Philosophy and science: The wide range of interaction.* Englewood Cliffs, NJ: Prentice Hall: 76–83.

Ehrlich, Paul R. 2000. *Human natures: Genes, cultures, and the human prospect.* Toronto: Penguin.

Expelled. 2008. Premise Media Corporation/Rampant Films. DVD.

Fagan, Brian. 2011. *Cro-Magnon: How the ice age gave birth to the first modern humans.* New York: Bloomsbury.

Finkelstein, Israel, and Neil Asher Silberman. 2002 (c.2001). *The Bible unearthed: Archaeology's new vision of ancient Israel and the origin of its sacred texts.* Toronto: Touchstone/Simon & Schuster.

———. 2006. *David and Solomon: In search of the Bible's sacred kings and the roots of the Western tradition.* Toronto: Free Press.

Foster, Charles. 2009. *The selfless gene: Living with God and Darwin.* Nashville: Thomas Nelson.
Gazzaniga, Michael S. 2008. *Human: The science behind what makes us unique.* New York: HarperCollins.
Hedges, Chris. 2009 (c.2008). *When atheism becomes religion: America's new fundamentalists.* Toronto: Free Press.
Hepper, Peter. Aug. 2005. "Unraveling our beginnings: On the embryonic science of fetal psychology." *The Psychologist* 18 (8): 474-77. British Psychological Society.
Holy Bible: New international version. 1978. Grand Rapids, MI: Zondervan.
Holy Bible: New revised standard version. 1989. Nashville: Thomas Nelson.
Hume, David. "Of personal identity." In Bowie, G. Lee, Meredith W. Michaels, and Robert C. Solomon, eds. 1996. *Twenty questions: An introduction to philosophy.* 3rd edn. Toronto: Harcourt Brace & Co.: 365–68.
Jacobs, A.J. 2007. *The year of living biblically: One man's humble quest to follow the Bible as literally as possible.* Toronto: Simon & Schuster.
James, William. "The stream of consciousness." In Ornstein, Robert E., ed. 1973. *The nature of human consciousness: A book of readings.* San Francisco: W.H. Freeman & Co.: 153–66.
Jethro Tull. 1974. *Warchild.* Chrysalis. LP.
Jung, Carl Gustav. "Synchronicity: An acausal connecting principle." In Ornstein, Robert E., ed. 1973. *The nature of human consciousness: A book of readings.* San Francisco: W.H. Freeman & Co.: 445–57.
Leiber, Justin. "How to build a person." In Bowie, G. Lee, Meredith W. Michaels, and Robert C. Solomon, eds. 1996. *Twenty questions: An introduction to philosophy.* 3rd edn. Toronto: Harcourt Brace & Co.: 357–64.
Locke, John. "Of identity and diversity." In Bowie, G. Lee, Meredith W. Michaels, and Robert C. Solomon, eds. 1996. *Twenty questions: An introduction to philosophy.* 3rd edn. Toronto: Harcourt Brace & Co.: 349–54.

Luce, Gay. "Biological rhythms." In Ornstein, Robert E., ed. 1973. *The nature of human consciousness: A book of readings*. San Francisco: W.H. Freeman & Co.: 421–44.

Lycan, William. "Robots and minds." In Bowie, G. Lee, Meredith W. Michaels, and Robert C. Solomon, eds. 1996. *Twenty questions: An introduction to philosophy*. 3rd edn. Toronto: Harcourt Brace & Co.: 201–07.

MacIntyre, Alasdair. "The story-telling animal." In Bowie, G. Lee, Meredith W. Michaels, and Robert C. Solomon, eds. 1996. *Twenty questions: An introduction to philosophy*. 3rd edn. Toronto: Harcourt Brace & Co.: 368–71.

Michaels, Meredith W. "Persons, brains, and bodies." In Bowie, G. Lee, Meredith W. Michaels, and Robert C. Solomon, eds. 1996. *Twenty questions: An introduction to philosophy*. 3rd edn. Toronto: Harcourt Brace & Co.: 354–56.

Morris, Simon Conway. 2003. *Life's solution: Inevitable humans in a lonely universe*. New York: Cambridge University Press.

Mosedale, Frederick E., ed. 1979. *Philosophy and science: The wide range of interaction*. Englewood Cliffs, NJ: Prentice Hall.

Ornstein, Robert E., ed. 1973. *The nature of human consciousness: A book of readings*. San Francisco: W.H. Freeman & Co.

Perry, John. "The first night." In Bowie, G. Lee, Meredith W. Michaels, and Robert C. Solomon, eds. 1996. *Twenty questions: An introduction to philosophy*. Toronto: Harcourt Brace & Co.: 332–49.

Pinker, Steven. "The mystery of consciousness." *Time Archive* January 19, 2007.

Plato. 1970. *The republic of Plato*. Translated with an introduction and notes by Francis MacDonald Cornford. New York: Oxford University Press.

Polkinghorne, John. 2002. *The God of hope and the end of the world*. New Haven: Yale University Press.

———. 2010. *The Polkinghorne reader: Science, faith and the search for meaning*. Thomas Jay Oord, ed. West Conshohocken, PA: Templeton Press.

Powell, Diane Hennacy. 2009. *The ESP enigma: The scientific case for psychic phenomena*. New York: Walker & Co.

Religulous. 2008. TVA Films/Thousand Words. DVD.

Robinson, James M. 2007. *The secrets of Judas: The story of the misunderstood disciple and his lost gospel.* Rev. edn. New York: HarperSanFrancisco.

Russell, Robert John, Nancey Murphy, and William R. Stoeger, S.J., eds. 2008. *Scientific perspectives on divine action: Twenty years of challenge and progress.* Berkeley, CA: The Center for Theology and the Natural Sciences.

Ryle, Gilbert. "The concept of mind." In Bowie, G. Lee, Meredith W. Michaels, and Robert C. Solomon, eds. 1996. *Twenty questions: An introduction to philosophy.* 3rd edn. Toronto: Harcourt Brace & Co.: 192–200.

Searle, John R. "The myth of the computer." In Bowie, G. Lee, Meredith W. Michaels, and Robert C. Solomon, eds. 1996. *Twenty questions: An introduction to philosophy.* 3rd edn. Toronto: Harcourt Brace & Co.: 207–13.

Stevenson, Leslie, ed. 1981. *The study of human nature.* New York: Oxford University Press.

———, and David L. Haberman. 2009. *Ten theories of human nature.* 5th edn. New York: Oxford University Press.

Stringer, Chris. 2012. *Lone survivors: How we came to be the only humans on earth.* New York: Times Books/Henry Holt & Co.

Tabor, James D. 2012. *Paul and Jesus: How the apostle transformed Christianity.* Toronto: Simon & Schuster.

Tiger, Lionel, and Michael McGuire. 2010. *God's brain.* Amherst, NY: Prometheus.

Wade, Nicholas. 2006. *Before the dawn: Recovering the lost history of our ancestors.* Toronto: Penguin.

———. 2009. *The faith instinct: How religion evolved and why it endures.* Toronto: Penguin.

Walter, Chip. 2008 (c.2006). *Thumbs, toes, and tears: And other traits that make us human.* New York: Walker & Co. (Macmillan).

Watson, Peter. 2005. *Ideas: A history of thought and invention from fire to Freud.* New York: HarperCollins.

White, L. Michael. 2010. *Scripting Jesus: The gospels in rewrite.* New York: HarperOne.

Wilson, Barrie. 2008. *How Jesus became Christian.* New York: St. Martin's Press.

Index

Number
2001, A Space Odyssey, 171

A
Abraham, 15, 16, 22
abstract,
abstract ideas
　ability to, 106
　capacity for, 134
Acts of Apostles, 16
Adam and Eve, 170
afterlife, 2, 31, 62, 77-78, 80, 91, 96, 141, 144, 193
aggression, organized, 98
AI, 71, 73, 75, 128. *See* artificial intelligence.
Alexander the Great, 115
Alzheimer's disease, 80, 85
amputation, effect of, 125
ancestor cult, 107
Anderson, Ian, 167
Animalia, 93
anthropic principle, 155
anthropology, 170
　discipline of, 93
anti-religionists, 162
apes, 93-94, 96
Appiah, Anthony, 85-87, 199
Aquinas, Thomas, 154
Aristotle, 40-42, 85, 91, 154
artificial intelligence (AI), 71, 73, 75, 128
ascension, 16
Assyrians, 7

atheists, 166-67
atman, 32, 89
atonement, 22
attachment, 34
Augustine, 48
australopithecine(s), 96, 99, 109
authors
　identities of, 28
　of Bible, 26
avatar, 34
awareness, tactile, 102

B
Babylonian captivity, 2
Babylonians, 6, 7
Bahamas-Atlantis connection, 159
Barbour, Ian, 174
baseball, 155-56
Beauregard, Mario, 135
behavior, 30, 46, 48, 51, 57-59, 71, 73, 75, 99, 121, 124, 129, 134-35, 139, 144, 158, 177, 187
　compulsive, 129
　observed of others, 99
behaviorism, 57
behaviorist psychology, 57
behaviorists, 58-59
Bhagavad Gita, 31
Bible, 1, 20, 61, 164
　as not airtight, 186
　authorship of, 18
　critiquing the, 28
　internal inconsistencies, 13
bibliolatry, 14

big bang, the, 169
biorhythms, 139
bipedality, 101, 109
birth, death and rebirth, cycle of, 32, 40
blackbody radiation, 152
Blue Dolphin, 155, 177, 204-05
body, as compound, 38
body and soul, 61
   dualism of, 43
bonobos, 94, 96, 100, 105, 134
Bowie, Michaels and Solomon, 61
Brahman, 32
brain(s), 44, 62-65, 68-70, 72, 74-77, 83-85, 88, 91, 95, 97, 101-03, 105-07, 109, 119, 121-30, 133-37, 139, 142, 180, 187-89, 199, 202-03
   as part of the body, 44
   big, 102
   complexity of, 128
   left hemisphere of, 126-27
   right hemisphere of, 126
   volume, 101
Buddha, 20, 33- 36, 41, 45, 171
   and Jesus, 35
   as transcendent, 33
Buddhism, 33-35, 58
Buddhist analyses, 77
Buddhists, 20, 36-37, 39, 48, 78, 89

C
Caesar, Julius, 18
Cambrian explosion, 170
Canaan, 1- 5, 177
   conquest of, 5, 177
   land of, 5
Canaanites, 2, 4
capitalism, 51-52
categorical imperative, 48
Catholicism, 163
cave analogy, of Plato. *See* Plato
Cayce, Edgar, 159
cerebral cortex, 125, 128
chemistry, 170
chimps, 96, 98-100, 103-106, 134
choices, 36, 48, 53-55, 112, 125, 179-80

Christianity, 33, 35, 38-40, 43, 145, 147-48, 165, 186, 203
   expressions of, 148
   history of, 3
Christians, 48
Christology, 185
chromosomes, 86
clairvoyance, 139, 142
Classics, the, 30
Clinton, Bill, 23
Collins, Francis, 161
communism, 51-52
compounds, 37
computer(s), 63, 71-75, 83, 91, 128, 203
computer programs, 73-74
conflict, 174-76
Confucianism, 31
Confucius, 29-31
conscience, social, 56
conscious, 78
consciousness, 58, 79-81, 83-84, 88, 96, 108, 110-11, 126, 130, 135, 182, 188-89, 195, 201-02
conservation of energy, 67-69, 131, 142
Conservatives, 146-47
constants of nature, physical, 37
contemplation, 41, 135
contingency, 72
conventional Christianity, 35
covenants, 49
Coyne, George, 161
craving
   and grasping, 34
   and Nirvana, 35
creationism, 151
creationists, 153
criticism
   Biblical, 120
   textual, 186
Cro-Magnons, 95
Crossan and Reed, 141-42
crying, 105
cyborg, 72-73
Cyrus, king of the Persians, 7

D

Daniel, book of, 2, 115
dark energy, 158, 178
dark matter, 158
Darwin, Charles, 2, 55-57, 93, 110, 148, 151-52, 154-56, 174, 187, 201
David, king, 17, 20, 22, 29, 45, 76, 178, 200-01, 203-05
Dawkins, Richard, 153
DDD, 189-90, *see* direct divine deposit
de Chardin, Pierre Teilhard, 177
death
 Jesus', 140
 of the brain, 137
 of the mind, 137
 point of, 62, 141, 142
decision-making, process of, 104
Decree of Heaven, 30
Descartes, Rene, 44-45, 62-65, 70-71, 76, 120, 200
design, theme of, 191
Destiny, 30
Determinism, 30
devil, Lucifer, or Satan, 9
dialogue, 174-76
direct divine deposit (DDD), 189-90
discipleship in Christ, 35
dishonesty, of Bible authors, 27
divinity, visibility of, 14
dogmatism, 119
Dowd, Michael, 166
dreams, 47, 88, 132-33, 139, 143, 193-95
dualism, 190
 and materialism, 68
dualistic interactionism, 37, 62-66, 68-70, 115-16, 124, 137-38, 140, 142, 190, 193
Ducasse, Curt John, 68-69, 118-19, 131, 133, 135-37, 140, 142, 200
duty, 48

E

Ecclesiastes, third chapter of, 173
education, 150
educational institutions, 151

Edwards, Paul, 66-68, 200
Egypt, 1, 2, 5, 114, 141
 exodus from, 5
Ehrman, Bart, 25-26
eightfold path. *See* middle way
Einstein, Albert, 139
electromagnetism, 140
elements, 18, 37-38, 117, 121, 160, 162, 169-70
Elijah, 18
eliminative materialism, 65
emotion, 104-05
 and intelligence, 126
empiricist, 83
enlightenment, 36
Enoch, 18
Epimenides of Crete, 185
epiphenomenalism, 65, 67-69
epiphenomenalist strategy, 66
epiphenomenalists, 68, 130
ESP, 142
ethology, 59
eugenics, 56
evangelicalism, 164
evil, 10, 48, 50, 56, 111, 115, 158, 173, 180, 182-84
evolution, 157, 167
 as God's ongoing creativity, 166
"Exceptional Human Experiences" or EHE, 138
Exodus, 177
"Expelled", 145, 150-51, 200
Ezra, 6-8, 120

F

faith, 160, 162
 and belief, 166
 secular, 163
fetus, life of the human, 121
Finkelstein and Silberman, 4
fire, harnessing power of, 96
Forms, realm of, 38-40, 44, 53, 178, 190-91
four noble truths. *See also* truths, four noble
free choice, 30
free will, 45-46, 48, 64, 69, 134

208 *God and the New Metaphysics*

freedom, 54
   academic, 150
Freud, Sigmund, 46-47, 53, 57-58, 71, 163, 203
functionalism, 66
functionalist argument, 73
fundamentalism, 157, 160
fundamentalists, 167
future, the, 162
fyborg, 72

G
Galileo, 174
Galton, Francis, 56
Gandhi, 25
Gazzaniga, Michael, 87, 96, 100, 102-06, 117, 122-23, 126, 128-29, 136, 193, 201
genes, 123
genetics, 62
genus, 155
"geo-physiology", 177
Germany, 4, 5, 57
Germ-line therapy, 87
God, 45, 79
   and humans, 178
   and Moses, 15
   and Satan, 9
   as catalyst, 171
   as designer, 155
   as lawgiver, 113
   as male, 16
   as possessing foreknowledge of events, 10
   as the bestower of life, 62
   breath of, 113
   cannot be seen, 14
   idea of, 49
   judgment of, 81
   messages to, 108
   misrepresenting, 6
   moral attributes of, 108
   of the three Abrahamic faiths, 42
   qualities of, 107
God's chosen people, 5
God's face, 15
God's image, 172
God's justice, belief in, 3

God's self, 32
God-consciousness, 111
gods, Greek and Roman, 11
God-substitutes, 112
Golden Rule, 30
gospel writers, 11-13, 21, 30, 146
gospels, 11-13, 16, 18-19, 26, 35, 192, 203
   Synoptic, 16
grasping, 35
Great Leap Forward, 96, 107
Greeks, 3, 32, 43, 115

H
Hamer, Dean, 135
Hanukkah, 115
happiness, 41
Haught, John, 174
heart(s), 107, 112
Hedges, Chris, 167
Hegel, 51
Hinayana. *See* Theravada
Hindu ideas, as different from Christianity, 33
Hinduism, 31-34, 87
Hindus, 48
Hitler, Adolf, 4
holy spirit, 142
hominids, 94
Homo erectus, 95, 101, 103, 106, 109
Homo ergaster, 95
Homo habilis, 95-96, 99
Homo heidelbergensis, 95
Homo sapiens, 95, 109, 167
human, 96
   as body and soul, 38
   skills, 100
   to be, 109
   vocal abilities, 101
human behavior, 57-58, 123
human condition, 3, 29, 34, 48, 54, 180
   graphing the (Figure 3), 181
human nature, 1, 6, 9, 14, 18-19, 23, 29-31, 43-44, 48, 52, 58, 61, 64, 75, 90-91, 117, 120-21, 123, 137, 145-46, 160, 172, 183, 191, 203
   and the devil, 10
   to shape information, 16

human potential, to follow the way
    of the sages, 30
human problem, basic, 36
humans, 14, 93
    and other animals, 97
    arrival of, 94
    as lying, 25
    as self-aware, 170
    composition of, 61
    identity problem, 32
    makeup of, 61
Hume, David, 44-45, 76, 89, 120-21, 201
Huxley, Thomas Henry, 65

I
"I", 87-88, 90-91,
icons, Catholic, 20
ID, 152, 155, 159-60, 171. *See*
    intelligent design
idealism, 66, 70
ideas, fear of, 149
identity, 89
identity theory, 65, 66, 68-70
image of God, 42, 108
imagination, 68
    topic of, 193
impermanence, 34, 36-37
improbability, 138
independence, 174-76
inner self, 32-33
instincts, 58-59
integration, 174-76
intelligence, 71, 106
    and emotion, 126
Intelligent Design (ID), 145, 151
intention, 17, 22, 58, 72, 123, 148, 171
Islam, 42, 145
Israel
    history of, 114
    kingdom of, 1, 2, 4, 5, 7-8, 113-15,
        141, 171, 200

J
Jackson, Michael, 86
Jacob, eye contact with God, 15
Jaspers, Karl, 171
Jesus, 16-17, 132
    and Buddha, 35-36
    and negative press, 24
    as conquering king, 20
    as exemplar, 25
    as God's anointed, 23
    as human, 116
    as sinless, 21
    as superhuman, 22
    ascension into heaven, 16
    death of, 182
    in the gospels, 12
    of Nazareth, 24
    of the text, 36
    person of, 185
    the risen, 165, 192
    untainted by sin, 21
Jesus Seminar, 12-13
Jews, 1- 3, 7, 18, 24, 42, 113-16, 172, 191
Job, 8-11, 15, 143
    book of, 9
Joseph, 20
Joshua, 3, 5-6
Josiah, 2
Judaism, 43, 115, 145
Judas, death of, 17
Judeo-Christian tradition, 42, 49, 90
Jung, Carl Gustav, 138

K
K'ung Fu-tzu, 29
Kant, Immanuel, 48-50, 53, 61, 120,
    149, 179
karma, 32, 34, 40
Kennedys, multiple, 18
Kelvin, Lord, 152

L
language, 105
    appearance of, 96
Lao-tzu, 171
Lazarus, 18
learning, procedural, 102
Leiber, Justin, 81-84, 201
lex talionis, 173
liberalism, 51
Liberals, 146-48
life, origin of on our planet, 153
Locke, John, 76, 79-81, 83-84, 89, 171,
    201

logical positivism, 121
Lorenz, Konrad, 59
Lotus Sutra, 33
Lovelock, James, 177
Luke, 11-12, 16, 20, 22, 24, 26, 37, 132, 140, 144, 165, 173
Luther, Martin, 50, 164, 185, 194
Lycan, William, 71-75, 202

M
Maccabean revolt, 2, 116
MacIntyre, Alasdair, 90-91, 126, 202
Mahayana, 33
Maher, Bill, 160-64, 166-68
Mammalia, 94
Mandelbrot set, 134
Mark, 12, 16, 19, 24, 26, 132, 140, 164
Marx, Karl, 51-55, 163
Mary, 21
materialism, 118, 157, 160
mathematics, 169
matter, 33, 188
Matthew, 12
meditation, 32, 89, 145
memory, 89, 136
  and mysticism, 135
  basis of, 179
  faulty, 80
  stores, 79
  theory, 76, 79
mental event, 63, 65
mental world, 46
mentality, 20, 46, 65-72, 76, 104, 110, 117, 120, 124, 129-30, 136-37, 145, 159, 174
Messiah, 23, 35, 43-44, 49-50, 146, 186
messianic expectation, 17
metaphysical, 190
metaphysics, 162
Michaels, Meredith W., 61, 66, 76, 79, 84, 199-03
middle way, of Buddhism, 34, 41
mind(s), 5, 12, 24, 27, 38, 41, 44-45, 47, 58, 61-84, 88, 91, 93, 98, 103, 107-08, 110, 113, 116, 120-24, 129-31, 137-39, 142, 148-50, 168, 183, 186, 189-90, 192, 200, 203

theory of (TOM). *See* theory of mind (TOM)
mirror self-recognition (MSR), 103
monism, 65
monistic physicalist camp, 66
monkeys, 93-94, 103
monotheism, ethical, 114
moral indignation, 100
moral perfection, path of, 30
morality, 49, 187
  and religion, 50
  cultivating a transcendent, 30
  human, 182
morphic resonance, 177
morphogenetic fields, 177
Moses, 14-15, 20, 22, 120, 165
Mother Teresa, 23, 55
Mount Sinai, 14, 165
music, 100
mystical experience, 135
mysticism, 135, 176
  and memory, 135

N
natural selection, 56-57, 96-97, 102, 110, 127, 133, 155, 157-58, 166, 181, 184, 187
nature
  of reality, 31
  our essential, 48
  science of, 117
NDE (Near Death Experiences), 138
Neanderthals, 95, 106, 109, 110
neural activity and religion, 134
neuroanatomy, 102
neurophysiology, 139
neuroplasticity, 124, 129, 180
neuroses, 46-47
New Testament, 26
Nirvana, 34-35
no self as such, 89
non-desire, 35

O
occasionalism, 69, 70
Old Testament, 27
One, the, 32, 39

one and the many, the, 32
origin of species, The, 57
original sin, doctrine of, 192
Overton, Judge William, 153

P
Paley, William, 154
Pali Canon, 33
parallelism, 69-70
parapsychology, 142
Paul, 13-14, 16, 26-28, 35, 50, 137-38, 143, 165, 185, 189, 190-92, 200, 203
    as author, 27
    letters of, 13, 27
Pentateuch, 120
perception
    in dreams, 133
    of physical objects, 133
Perry, John, 77
Persia, 115
person, essence of a, 86
personal identity, 61, 75-76, 79, 81, 84, 86-87, 89-91, 191, 201
personalities, split, 78
personality, 78, 107, 151
personhood, 72, 75-76, 80, 83, 88-89
perspective, topic of, 185
Peter, as not an author, 27
Pharisees, 12
photoelectric effect, 152
phyla, 170
physicalism, 65, 70
physicality, 68
pineal gland, 63
Pinker, Steven, 119
placebo effect, 69, 131
Planck, Max, 152
"planetary medicine", 177
Plato, 37-40, 43-44, 46-47, 53, 62, 64, 77, 137, 191, 202
Plato's *Republic*, 39
Polkinghorne, John, 141
Pope, the, 163
prayer, 108, 131
precognition, 142
predetermination, genetic, 124

pre-established harmony, 69
prefrontal cortex, 103
Primates, 94
private property, 107
process theology, 185
processes, automatic, 123
prophecy, 19
Protagoras, 171
psychetricians, 82
psychology
    "science" of, 120
    evolutionary, 158

Q
Q source, 12, 13
quantum theories, 152
quantum world, 131

R
race
    and gender, 86
    importance of, 87
Rand, Ayn, 183
rationalist, 83
real, the truly, 38
reality, nature of, 31, 37
reason, 39
    pure practical, 48
rebirth, 34
reciprocity, lack of, 100
regression hypnotherapy, 84
reincarnation, 32, 34, 38-39, 81, 83-84, 140
relativity, 152
    theory of general, 139
religion, 49, 160, 166
    and neural activity, 134
    and science, 174
    hallmark of, 161
    topic of, 134
religion and science, ways to relate (Figure 2), 176
"Religulous", 145, 160
reproduction, 188
resurrection, 3, 16, 44, 77, 79, 81, 83-84, 116, 141, 143-44, 165, 189, 191-93

resurrection *(continued)*
    Jesus' own, 143
    of the dead, 3
Reuben, 25
rewards and punishments, system of, 49
Rig Vedas, 31
right desire, 35
Rolling Stones, 34
Roman Catholic Church, 56
Rousseau, 49
Ryle, Gilbert, 70, 71, 73-75, 203

S
salvation, 32
Sartre, Jean-Paul, 53-55, 59
Satan, as deceiver, 10
Saul, 20, 50, 165
science, 64, 117
    and religion, 174
    history of, 118
    limitations of, 119
    nature of, 117
sciences,"hard and soft", 169
scientific revolution, 120
scriptures
    Hindu, 31
    Judeo-Christian, 14
Searle, John R., 73-75, 203
self, 89, 128
    "essential", 192
    ego, 32
    inner, 32
    nature of this, 106
    notion of a, 187
    sense of, 105
    true, 32
Self
    God's, 33
    ultimate, 32
self-hood, 88
self-identity, 85
self-interest, 172
self-narration, 126
semantics, 74, 105
senses, 143
sentence structure, 105

Sermon on the Mount and the Plain, 173
Sheldrake, Rupert, 124, 177, 180, 205
Sheol, 2, 115-16, 141, 143, 191-92
Siddhartha Gautama, 33
silver cord, 141, 143
Simeon and Levi, vengeful acts, 25
Simon of Cyrene, 16
singing, 100
Skinner, B. F., 58-59
social insects, 97
Solomon, 25, 66, 199-203
Somatic gene therapy, 87
somatic theory, 84-85
somaticians, 82
soul, 31, 34, 38-41, 43-44, 61-62, 64, 69, 76-81, 83, 91, 107-10, 113, 116, 137, 141, 170, 189-91, 199
    able to survive bodily death, 62
    Aristotle's view of, 40
    as element, 38-39
    as "tripartite", 39
    Descartes', 44
    Plato's, 44
    within a body, 62
space, 19, 63, 67, 70, 120, 139-40, 169
species, 94, 155
speech, 96, 100, 117, 151-52, 194
Spencer, Herbert, 56
spirit, 33, 39, 91, 173, 188
    body without, 113
    release of the, 142
Stalin, Josef, 5
Stein, 151, 153
Stevenson, Leslie, 29
Stevenson and Haberman, 29, 31, 33, 35, 39, 44, 46, 49, 52, 55, 57, 137
subjectivity, 123
suffering, human disease of, 36
superstring theory, 178
symbol systems. *See* language
symbols, 74
synchronicity, 138
syntax, 105

T
tabula rasa, 83, 171

tactical deception, 104
telekinesis, 63-64, 142, 189
teleological, 154
telepathy, 139, 142
Ten Commandments, 11, 14, 172
theology, discipline of, 163
theory of mind (TOM), 103, 104, 106, 117, 139
therapist, task of, 47
Theravada. *See* Hinayana
thought, abstract, 105
thumbs
    flexible, 99
    opposable, 95, 98
TOM. *See* mind, theory of
Torah, 27-28, 113
total recall, 79
truth, 119
truths, four noble, in Buddhism, 34, 35
Tull, Jethro, 167
Turing, Alan, 73
Turing test, 73, 74
twins, genetic sameness of, 122

U
ulnar opposition, 99
Upanishads, 31-32, 171

V
van Lommel, Pim, 138
vegetarianism, 94
Vienna Circle, 121
virtue, 41

W
Wallace, Alfred Russell, 110, 189
Walter, Chip, 99
warfare, 98, 129
    spiritual, 129
Watson, J. B., 58, 116, 120-21, 128, 203
White, L. Michael, 13
White, Rhea, 138
Wilson, E.O., 59
Wisdom, John, 158
worldview, Hindu, 32

Y
yoga, 32

Z
Zoroaster, 171
Zoroastrians (-ism), 114-115

About the Author

HERB GRUNING, PH.D., has taught a variety of courses in Religious Studies at a number of universities in both Canada and the United States, including McMaster University in Hamilton, Ontario, Canada. He most recently taught at Canisius College, a Jesuit institution in Buffalo, New York. His courses include Religion and Science, and Christian Concepts of God. His main area of research is the thought of the philosopher-physicists Alfred North Whitehead and David Bohm.

Dr. Gruning completed his Ph.D. from McGill University in Montreal, Quebec, Canada, in the area of Philosophy of Religion. He graduated from the program with honors (cum laude), and his dissertation was published by the University Press of America under the title, *How in the World Does God Act?* (2000). Blue Dolphin published his second book, *God & the New Metaphysics*, in 2005 as well as his third, *God Only Knows*, in 2009.

Born in Toronto, Ontario, Canada, Dr. Gruning, and his wife Alice, now live in London.

More Books from Blue Dolphin Publishing

God Only Knows
Piecing Together the Divine Puzzle
Herb Gruning, Ph.D.
ISBN: 978-1-57733-240-4, 173 pp., 6 x 9, $16.95

Of the many ways of thinking about God, three models stand out: the traditional approach has the most seniority, but that alone does not give it security; the process scheme lies at the opposite extreme, but is too radical for some; and the newest option called "open" or "free-will" theism resides somewhere in the middle. "You can't tell the players without a program," but with one, a selection might come easier. This book is that program.

God and the New Metaphysics
Herb Gruning, Ph.D.
ISBN: 978-1-57733-161-2, 212 pp., 6 x 9, paper, $16.95

Examines metaphysical and cosmological proposals for an alternate vision of reality. It engages the thought of some who begin to disentangle us from the metaphysical knots in which we have tied ourselves, such as Alfred North Whitehead, James Lovelock, Rupert Sheldrake, Pierre Teilhard de Chardin, and David Bohm.

The Fifth Gospel
New Evidence from the Tibetan, Sanskrit, Arabic, Persian and Urdu Sources About the Historical Life of Jesus Christ After the Crucifixion
Fida Hassnain & Dahan Levi
ISBN: 978-1-57733-181-0, 344 pp., 6x9, paper, $19.95

Orders: 800-643-0765 • www.bluedolphinpublishing.com